U.S.-Russian COOPERATION IN SPACE

**Office of Technology Assessment
Congress of the United States**

Photo Credit: National Aeronautics and Space Administration

Recommended Citation: U.S. Congress, Office of Technology Assessment, *U.S.-Russian Cooperation in Space,* OTA-ISS-618 (Washington, DC: U.S. Government Printing Office, April 1995).

For sale by the U.S. Government Printing Office
Superintendent of Documents, Mail Stop: SSOP, Washington, DC 20402-9328
ISBN 0-16-048019-1

Foreword

The recent broad political rapprochement between the United States and the nations of the Former Soviet Union (FSU) has transformed the environment for cooperation on space projects, and led to cooperative programs in space with Russia and other FSU states that would have been unimaginable just a few years ago. Chief among these are the high-profile human spaceflight cooperative activities involving the Space Shuttle-Space Station Mir dockings and the International Space Station.

This report surveys the potential benefits and drawbacks of expanded cooperation with Russia and other nations of the FSU in space activities, and examines the impacts of closer cooperation on U.S. industry and U.S. national security concerns. Such cooperation has begun to yield scientific, technological, political, and economic benefits to the United States. However, the political and economic risks of cooperating with the Russians are higher than with the United States' traditional partners in space. Cooperation in robotic space science and earth remote sensing is proceeding well, within the stringent limits of current Russian (and U.S.) space budgets. Including Russia in the International Space Station program provides technical and political benefits to the space station partners, but placing the Russian contribution in the critical path to completion also poses programmatic and political risks.

The report notes that much of the motivation for the expansion of cooperation with Russia lies beyond programmatic considerations. In particular, it points out that continued cooperation, including large payments for Russian space goods and services, may help stabilize Russia's economy and provide incentive for some of Russia's technological elite to stay in Russia and contribute to peaceful activities in space. Lack of opportunities at home might otherwise cause them to seek employment abroad where their skills might contribute to the proliferation of weapons of mass destruction. Finally, the report assesses the pros and cons of expanded commercial ties, their impact on the U.S. space industrial base, and on aerospace employment.

In undertaking this effort, OTA sought the contributions of a wide spectrum of knowledgeable individuals and organizations. Some provided information; others reviewed drafts. OTA gratefully acknowledges their contributions of time and intellectual effort.

ROGER C. HERDMAN
Director

Workshop

Issues in U.S.-Russian Cooperation in Space
November 9, 1994

Kenneth S. Pedersen
Chairman
Research Professor of
  International Affairs
Georgetown University

Alain Dupas
Advisor to the President
CNES

Lewis R. Franklin
International Security Consultant

Louis Friedman
Executive Director
The Planetary Society

James W. Head, III
James Manning Professor
Brown University

Carolyn L. Huntoon
Director
NASA Johnson Space Center

Wesley Huntress
Associate Administrator for Space
  Science
National Aeronautics and Space
  Administration

Nicholas L. Johnson
Principal Scientist
Kaman Science Corporation

Charles F. Kennel
Associate Administrator
Mission to Planet Earth
National Aeronautics and Space
  Administration

E. William Land, Jr.
Principal
ANSER

Grant Lichtman
Director
WorldMap International, Ltd.

Charles H. Lloyd
President
LKE International

John Logsdon
Director
Space Policy Institute
George Washington University

Jeffrey Manber
Managing Director
American Operations
NPO Energia Ltd.

Marcia Smith
Specialist in Aerospace Policy
Library of Congress

Judyth Twigg
Instructor
Virginia Commonwealth
  University

Note: OTA appreciates and is grateful for the valuable assistance and thoughtful critiques provided by the workshop participants. The participants do not, however, necessarily approve, disapprove, or endorse this report. OTA assumes full responsibility for the report and the accuracy of its contents.

Project Staff

Peter Blair
Assistant Director, OTA
Energy, Materials and
   International Security Division

Alan Shaw
Director
International Security and Space
   Program

Ray Williamson
Project Director

Peter Smith[1]
Mark Suskin

CONTRACTORS
Cynthia Allen
Madeline Gross

ADMINISTRATIVE STAFF
Jacqueline R. Boykin
Don Gallagher
N. Ellis Lewis

---

[1]On detail from the National Aeronautics and Space Administration.

Acknowledgments

This report has benefited from the advice of many individuals. In addition to members of the workshop, the Office of Technology Assessment especially would like to thank the following individuals for their assistance and support. The views expressed in this paper, however, are the sole responsibility of OTA.

Paul Bowersox
Department of Political Science
University of Colorado

Kent Bress
Space Flight Division
Office of External Relations
National Aeronautics and Space
    Administration

Elizabeth Carter
International Relations Division
Office of External Relations
National Aeronautics and Space
    Administration

Leslie Charles
Mission to Planet Earth Division
Office of External Relations
National Aeronautics and Space
    Administration

Lynn F. H. Cline
Space Flight Division
Office of External Relations
National Aeronautics and Space
    Administration

Marc Constantine
Aerojet Corporation

Charles T. Force
Office of Space Communications
National Aeronautics and Space
    Administration

Joseph P. Loftus
Johnson Space Center
National Aeronautics and Space
    Administration

Patricia Maliga
Space Flight Division
Office of External Relations
National Aeronautics and Space
    Administration

J. Donald Miller
International Relations Division
Office of External Relations
National Aeronautics and Space
    Administration

P. Diane Rausch
Space Flight Division
Office of External Relations
National Aeronautics and Space
    Administration

Lisa Shaffer
Mission to Planet Earth Division
Office of External Relations
National Aeronautics and Space
    Administration

The project staff would also like to thank the following OTA staff for their review of the drafts of this report.

Richard Brody

Anthony Fainberg

S. Yousef Hashimi

Tom Karas

Contents

Executive Summary

The end of the Cold War, the collapse of the Soviet Union, and the changing world order have provided new opportunities and new incentives for the United States and other countries to cooperate with Russia in space science, space applications, and human spaceflight. Although U.S. attempts to cooperate on space activities with the Soviet Union began more than 30 years ago, intense political and military competition between the two countries severely limited the scope and duration of such activities. Today, the United States government is actively pursuing cooperation with Russia on a wide range of space activities, including the International Space Station. In addition, U.S. aerospace firms have entered into joint ventures, licensing agreements, and cooperative technical agreements with a variety of newly organized Russian counterparts.

The emergence of Russia as a major cooperative partner for the United States and other spacefaring nations offers the potential for a significant increase in the world's collective space capabilities. Expanding U.S.-Russian cooperation in space since 1991 has begun to return scientific, technological, political, and economic benefits to the United States. Yet, Russia is experiencing severe economic hardship and its space program has undergone major structural changes. The future success of U.S.-Russian cooperative projects in space will depend on:

- successful management of complex, large-scale bilateral and multilateral cooperative projects;
- progress in stabilizing Russia's political and economic institutions;
- preservation of the viability of Russian space enterprises;
- flexibility in managing cultural and institutional differences;

| 1

- continued Russian adherence to missile-technology-proliferation controls; and
- additional progress in liberalizing U.S. and Russian laws and regulations in export control, customs, and finance.

FOREIGN POLICY BENEFITS AND RISKS

Russia's technical contributions to the International Space Station offer a substantial increase in planned space station capabilities. Just as important to the United States are the foreign policy gains from this and other human spaceflight projects, such as the Shuttle-Mir dockings. U.S. officials expect cooperative activities to help promote economic and political stability in Russia. For example, the National Aeronautics and Space Administration's (NASA's) purchase of nearly $650 million in goods and services from Russia during fiscal years 1994-97, by far the largest transfer of U.S. public funds to the Russian government and private organizations, is an important signal of U.S. support for Russia's transition to a market economy. These purchases should help preserve employment for Russian engineers and technicians in at least some of Russia's major space-industrial centers, thereby inhibiting proliferation through "brain drain" and helping to sustain Russian adherence to the Missile Technology Control Regime. Moreover, NASA's purchases improve the chances that Russia will be able to meet its obligations to the space station project, thereby enhancing prospects for success.

Nevertheless, such purchases entail some political risk in the United States, as well as the risk to the space station if the Russian government and enterprises are not able to perform. Some U.S. observers question the wisdom of supporting any part of the Russian aerospace industry, which provided much of the technological substance for the Soviet threat to the United States; others believe that U.S. officials have made adequate provision to ensure that U.S. funds remain in the civil space sector.

OTHER BENEFITS AND RISKS

NASA is exploring cooperative space research and development with Russia in virtually every programmatic area. Aside from the space station, activities include flights of instruments on each other's spacecraft and joint missions using Russian launch capabilities with U.S.-built spacecraft. Public sector cooperation in space science and Earth observations is developing well for the most part. The political, technical, and administrative risks involved are somewhat higher than they are in NASA's traditional cooperative relationships, but—except for the space station—Russian contributions are not in the "critical path" to completion of key projects; program managers understand the risks involved and have made contingency plans to minimize long-term risks.

Cooperation on projects involving human spaceflight involves both potentially greater programmatic benefits and higher risks than it does in space science and applications. The United States stands to gain new experience in long-duration spaceflight and a better understanding of Russia's technology and methods. On the other hand, the United States risks possible project failure if Russia proves unable to perform as promised.

Placing the Russian contribution in the critical path to completion of the space station poses unprecedented programmatic and political risks. The Russian elements must be delivered on time and within budget; failure to do so could cause serious difficulties, both programmatically and in NASA's relations with its other partners and with Congress. Knowledgeable observers express concern about the stability and staying power of the Russian aerospace sector, about the Russian track record in delivering new spacecraft, and about the condition of the Baikonur launch complex (used to launch Proton and Soyuz vehicles). On another level, observers worry that political and/or military events within Russia or between Russia and other countries could cause either party to seek to amend the space station program or withdraw from it.

Given the significance of the Russian contribution to the space station, the U.S. ability to make up for delays or failure to deliver is severely limited by available U.S. resources. However, participants in current cooperative ventures suggest some other precautions that could be taken, both in the space station project and in space robotic cooperation:

- Seek better understanding of the larger political and economic forces that could affect Russian ability to deliver on commitments, perhaps through further systematic analysis of Russian aerospace industry developments.
- Maximize open and frank communication. To avoid as many technical and managerial surprises as possible, seek (and be willing to allow) a high degree of communication and interpenetration between the U.S. and Russian programs.
- Be prepared for delays and reverses.
- Be aware of and manage cultural differences effectively.

COMMERCIAL COOPERATION

Because of the potential for diverting civilian space technologies to enhance Soviet military capabilities, during the Cold War, the federal government effectively precluded U.S. aerospace firms from entering into cooperative business agreements with Russian entities. Now, most large U.S. aerospace companies are pursuing some form of joint venture or partnership with Russian concerns, especially in launch services and propulsion technologies. Although several of these emerging commercial partnerships show promise, and some could result in large revenues, none of them yet appear to be profitable, and it is too early to tell how successful they will be. Here, too, the risks are larger than they are in cooperative ventures with Japanese and Western aerospace firms because of unstable Russian political, economic, and legal conditions and potential linkage to U.S.-Russian political relations. The U.S. government could assist U.S. industry by further liberalizing U.S. export-control laws and regulations.

RUSSIA, THIRD PARTIES, AND THE UNITED STATES

The French experience in cooperating with the Soviet Union and Russia since 1966 largely parallels and confirms that of the United States. The European Space Agency has budgeted over $320 million for space cooperation with Russia, largely for European-built hardware that will be installed in the Russian portion of the International Space Station.

The U.S. decision to bring Russia into the space station partnership initially caused considerable strain in relations with the existing partners, already frayed by years of U.S. design changes and cost increases and aggravated by a general cooling of public enthusiasm for human spaceflight. Challenging negotiations remain to complete the realignment of the agreements covering the station's construction and utilization, but the working relationships now appear to be developing more smoothly.

DOMESTIC ECONOMIC IMPACT

Experts disagree over the nature and extent of the effect that expanded cooperation with Russia will have on the U.S. aerospace industry, and particularly on the retention of U.S. jobs. Some industry officials have expressed concern that U.S. aerospace employment could be lost and the technological base adversely affected by use of Russian technology in the U.S. space program. Others have argued that skillful incorporation of Russian technologies into U.S. projects could save taxpayer dollars in publicly funded programs such as the space station and could boost U.S. international competitiveness in commercial programs. Both could happen and have to be weighed against each other.

Russian launch vehicles and related systems have the most obvious potential for U.S. commercial use, but using them could adversely affect the U.S. launch industry. This industry is the subject of upcoming OTA reports.

Introduction and Findings | 1

INTRODUCTION

The U.S. civilian space program began in large part as a competitive response to the space accomplishments of the Soviet Union. For the first three decades, at least, competitive impulses played a major role in the direction of U.S. space activities. Cooperation with the Soviet Union was highly limited, with the most important projects being undertaken in an attempt to open lines of political communication between the two superpowers.[1] The recent collapse of the Soviet Union and the end of the Cold War have brought dramatic changes to the civilian space programs of both the United States and the Former Soviet Union (FSU). Once implacable adversaries who used their space programs to demonstrate scientific and technical prowess, the United States and the countries of the FSU (figure 1-1) are now seeking to develop a variety of political, economic, and other ties to replace their Cold War competition.

U.S.-Russian cooperation on the International Space Station, begun in 1993, is the largest and most visible sign of the new relationship in space activities. However, the United States and Russia have embarked on a host of other cooperative space projects involving both government and industry. These ventures range from administratively simple projects between individual scientists to complicated commercial and intergovernmental transactions. Such activities also involve a wide range of investments, from a few thousand to hundreds of millions of dollars. Although

---

[1] See, e.g., U.S. Congress, Office of Technology Assessment, *U.S.-Soviet Cooperation in Space*, OTA-TM-STI-27 (Washington, DC: U.S. Government Printing Office, July 1985).

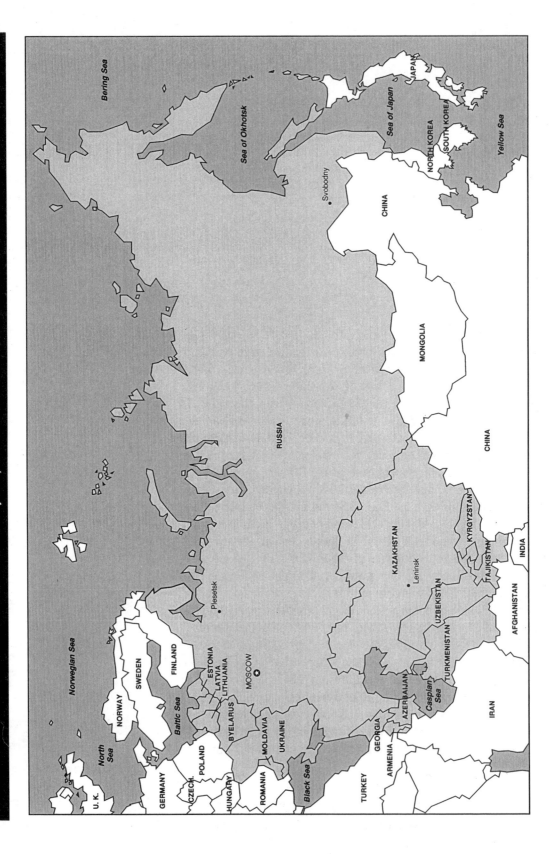

FIGURE 1-1: Newly Independent States of the Former Soviet Union

the United States' most extensive partnerships with FSU countries are with the Russian government and Russian commercial entities, the United States and U.S. companies are exploring additional cooperative ventures with other FSU countries.

This rapid expansion of cooperative activity is taking place in the context of serious economic decline, political instability, and social disruption in Russia and the other countries of the FSU. As a result, both the potential benefits and the risks involved are considerably greater than they are in U.S. space cooperation with Europe, Japan, and other partners. This is particularly the case for the International Space Station program. By engaging the Russians as partners and purchasing related Russian space hardware and services, the United States hopes to benefit not merely from Russian technical capabilities, but from improved Russian political and economic stability and continued adherence to nonproliferation goals. At the same time, Russian failure to deliver as promised, for whatever reason, could risk the future of the program itself.

This report surveys issues related to U.S.-Russian efforts—governmental and commercial—to cooperate in civil space activities.[2] It was requested by the House Committee on Science, which asked the Office of Technology Assessment (OTA) to "undertake an initial survey of issues related to U.S.-Russian cooperation in space activities."[3] The committee also asked OTA to extend its analysis to other republics of the FSU, where applicable.

To gather data for this report, OTA convened a workshop of experts with experience in U.S.-Russian cooperative efforts in space science, space applications, space-launch services, and human spaceflight.[4] The workshop gave OTA the opportunity to explore the lessons learned in previous or ongoing U.S.-Russian cooperative programs and to discuss the implications for future cooperative efforts.

Chapter 1 presents OTA's major findings regarding U.S.-Russian cooperation. Chapter 2 summarizes the status and organization of the Russian space program and shows how it relates to programs of other FSU countries. The history and current state of U.S.-Russian cooperation are explored in chapter 3. Chapter 4 summarizes lessons learned from others who have developed major space ties with Russia and its predecessor, as well as past and potential interactions among Russia's partners and the United States. Chapter 5 examines risk management, the role of governments, and opportunities for and impediments to establishing or expanding cooperative relationships. Finally, chapter 6 examines the impacts of closer cooperation on U.S. industrial and national security concerns.

FINDINGS

FINDING 1: *The dramatic expansion of U.S.-Russian cooperation in space since 1991 has begun to return scientific, technological, political, and economic benefits to the United States. Further cooperative gains will depend on:*

- successful management of extraordinarily complex, large-scale bilateral and multilateral cooperative projects;
- progress in stabilizing Russia's political and economic institutions;
- preservation of the viability of Russian space enterprises;
- successful management of cultural and institutional differences;
- continued Russian adherence to missile-technology-proliferation controls; and

---

[2] This report deals only with civil space cooperation and does not address cooperative activities of the military or of the Department of Defense.

[3] George E. Brown, Jr., and Robert S. Walker, House Committee on Science, Space, and Technology, letter to Roger Herdman, Director of OTA, Aug. 29, 1994.

[4] Held Nov. 9, 1994, in Washington, DC.

BOX 1-1: General Benefits of Cooperation in Space

- **Reducing costs and sharing burdens**. Many of the agencies involved in space share common goals and have developed overlapping programs. Facing budget constraints, these agencies are looking for ways to coordinate their programs to eliminate unnecessary duplication and to share the cost burden of projects they might otherwise do on their own.
- **Broadening sources of know-how and expertise.** Scientists and engineers from other countries may possess technology or know-how that would improve the chance of project success.
- **Increasing effectiveness.** The elimination of unnecessary duplication can also free up resources and allow individual agencies to match their resources more effectively with their plans. This reallocation of resources can eliminate gaps that would occur if agency programs were not coordinated. International discussions can be valuable even if they merely help to identify such gaps, but they can be particularly useful if they lead to a division of labor that reduces those gaps.
- **Aggregating resources for large projects.** International cooperation can also provide the means to pay for new programs and projects that individual agencies cannot afford on their own. This has been the case in Europe, where the formation of the European Space Agency has allowed European countries to pursue much more ambitious and coherent programs than any of them could have accomplished alone.
- **Promoting foreign policy objectives.** Cooperation in space also serves important foreign policy objectives, as exemplified by the International Space Station program. The agreements on space cooperation reached in 1993 and 1994 by Vice President Al Gore and Russian Prime Minister Viktor Chernomyrdin have also led to significant cooperative activities in space science and Earth observations.

SOURCE: Office of Technology Assessment, 1995.

- additional progress in liberalizing U.S. and Russian laws and regulations in such areas as export control, customs, and finance.

Most knowledgeable observers conclude that international cooperation will be essential to the success of future major space plans.[5] Most large programs are too ambitious to be undertaken as unilateral efforts.[6] Multilateral projects and extensive coordination among all potential partners in a particular field are becoming increasingly common as all major space-faring nations encounter significant budget pressures yet desire to

accomplish more in space (box 1-1). The emergence of Russia as a major cooperative partner for the United States and other space-faring nations offers the potential for a significant increase in the world's collective space capabilities.

U.S.-Russian cooperation in space is taking a wide variety of forms, ranging from relatively straightforward company-to-company arrangements to high-profile government-to-government cooperative agreements. Most large U.S. aerospace companies are pursuing some form of joint venture or partnership with Russian concerns. The

---

[5] Kenneth Pedersen, "Thoughts on International Cooperation and Interests in the Post-Cold War World," *Space Policy* 8(3): 205-220, August 1992; George van Reeth and Kevin Madders, "Reflections on the Quest for International Cooperation," *Space Policy* 8(3): 221-232, August 1992; American Institute of Aeronautics and Astronautics (AIAA), *Conference on International Cooperation* (Kona, HI: AIAA, 1993); U.S.-CREST, *Partners in Space* (Arlington, VA: U.S.-CREST, May 1993); John M. Logsdon, "Charting a Course for Cooperation in Space," *Issues in Science and Technology*, 10(1): 65-72, fall 1993.

[6] U.S. Congress, Office of Technology Assessment, *International Collaboration in Large Science Projects* (Washington, DC: U.S. Government Printing Office), forthcoming, Spring 1995.

National Aeronautics and Space Administration (NASA) is exploring cooperative space research and development (R&D) in virtually every area that interests its scientists and engineers. Intensified cooperation with Russia, either bilaterally or in a multilateral framework, could yield great benefit for both countries.

Cooperation between the two countries raises economic, financial, scientific, foreign policy, and national security issues. U.S. efforts to include Russian scientists and engineers in cooperative efforts derive in large part from a desire to help Russia make a successful, stable transition to democracy, develop a market economy, and reduce military production in favor of civilian manufacturing. By involving some portion of Russia's technical elite in high-technology space projects, the United States hopes to encourage highly educated professionals to stay in Russia and help develop its economy, rather than move to countries potentially hostile to the United States and its allies.[7] U.S. purchases of space-related goods and services from Russia also provide much-needed hard currency for the Russian economy. Cooperation on Earth-observation projects stems in part from a desire to involve Russia more deeply in regional and global environmental matters.

The recent broad political rapprochement between the United States and Russia has transformed the environment for cooperation on space projects. Previously, both governments limited what could be done, for political reasons and because of the desire to prevent the transfer of strategically useful technical information; conversely, efforts to ease political tensions occasionally stimulated the pursuit of cooperative activities that might not otherwise have been considered of high scientific or engineering value, such as the Apollo-Soyuz Test Project (see photo on page 10).[8] NASA program managers constantly faced the reality that the political linkage—that is, the linkage between politics and support for certain projects—could work to disrupt cooperative undertakings, as events in the Soviet Union, Afghanistan, and Poland did at the end of the 1970s and early 1980s.[9] Today, the desire to support economic and political stability in Russia and to provide tangible incentives for positive Russian behavior in areas such as nonproliferation of weapons of mass destruction and their delivery systems encourages cooperation.[10] As a result, the United States has made unprecedented commitments of resources to cooperative projects, including purchases of Russian goods and services, and has been willing to make Russian hardware and launch services major components of keystone NASA projects, particularly the International Space Station.

Technologically, Russian hardware and other capabilities have much to offer the space station and other projects. To learn more about working together and to gain early long-duration-flight experience, NASA has embarked on a two-year series of engineering and scientific experiments involving the Mir Space Station and the U.S. Space Shuttle.[11] Nevertheless, more intense cooperation entails some significant risks and liabilities. Political and economic instabilities with-

---

[7] U.S. Congress, Office of Technology Assessment, *Proliferation and the Former Soviet Union,* OTA-ISS-605 (Washington, DC: U.S. Government Printing Office, September 1994).

[8] The Apollo-Soyuz Test Project involved an orbital rendezvous between a U.S. Apollo capsule and a Soviet Soyuz capsule. The project was planned and carried out during the early 1970s (see ch. 3 for greater detail).

[9] Human rights abuses in the Soviet Union, the Soviet invasion of Afghanistan in 1979, and the institution of martial law in Poland in 1981 occasioned a sharp cooling in the cooperative relationship.

[10] U.S. Congress, Office of Technology Assessment, *Technologies Underlying Weapons of Mass Destruction,* OTA-BP-ISC-115 (Washington, DC: U.S. Government Printing Office, December 1993).

[11] This arrangement gives the United States access to long-duration-flight opportunities for the first time since the mid-1970s, when the United States launched and occupied Skylab.

NATIONAL AERONAUTICS AND SPACE ADMINISTRATION

Apollo-Soyuz Test Project.

in Russia constitute the greatest risks to the pursuit of cooperative activities. After investing both time and money in cooperative programs, if Russia failed for any reason to follow through, the United States could be faced with having to complete such programs on its own, cancel them, or find new partners. Companies face the risk of losing their investment of time and money if commercial agreements fail. In time, companies also risk the loss of entire programs or product lines.

Russia has undergone enormous political changes, having gone from a centralized regime under the Soviet flag to an emerging democracy in just a few years. Its newly formed democratic institutions are still quite fragile.[12] Russia is also attempting to move from a centrally planned economy to one in which market forces predominate. Such changes have imposed considerable hardship on Russia's people. Rapid political and economic improvements are impeded by the very human impulse to resist change. To solidify changes in its political and economic order, Russia must, therefore, build the legal and commercial infrastructure to support and enhance the changes. Because the political, economic, and administrative nature of Russian private and governmental institutions is changing rapidly, each cooperative agreement generally requires charting new institutional ground, adding to the uncertainties of cooperating with Russia.

Russians and Americans have strong cultural differences. Over the nearly 40 years of mutual isolation in technical matters, the two countries have also acquired different technical and managerial approaches to the development and application of space technology. Such differences can be beneficial to both sides because they add new perspectives, but they can also be barriers to increased cooperation. By contrast, the European, Japanese, and Canadian space communities have had a close relationship with their U.S. counterpart, making communication and collaboration much easier than they are between the United States and Russia. Although cooperation with Canada, Europe, and Japan has its own set of risks, cooperation with Russia is currently more difficult. Workshop participants pointed out the importance of maintaining open minds and learning more about Russian practices in order to reduce misunderstandings resulting from cultural differences. The uncertainties of cooperating with Russia (box 1-2) suggest that the U.S. government and

---

[12] The debate within Russia over the Russian military's recent attempts to prevent the Russian Republic of Chechnya from seceding from Russia underscores the vulnerability of these new institutions.

BOX 1-2: Uncertainties of Cooperating with Russia

- **Technical risks.** Despite Russia's prowess in developing and maintaining a large and capable space program, it has certain weaknesses, such as difficulty maintaining schedules on new spacecraft and components, which were evident even before the end of the Cold War. Russia will have to complete several new systems to fulfill its upcoming cooperative and contractual commitments.
- **Unstable political institutions.** Russian democratic institutions are in a very early stage of development, and successful maturation is far from certain. Legal and political instability is great and appears likely to remain so for some time to come.
- **Russian military actions.** The Russian military has undergone substantial change in the past few years and is much less stable than it was under the U.S.S.R. government. Instability in the Russian military could make the Western world much more wary about investing in Russia and could even undermine economic and political stability. For example, the war in Chechnya has drained important resources from the civilian economy and has raised concerns about human rights abuses.
- **Economic uncertainty.** The near collapse of the Russian economy and its impact on the many enterprises essential to Russian space activity could affect Russia's ability to deliver on international commitments. Russia lacks a common, settled business and procedural framework within which to organize and regulate its new marketplace.
- **Crime and corruption.** The political and legal changes in Russia and lax enforcement have increased the incidence of serious crime and open corruption, thus impeding the development of normal business relationships.
- **Cultural barriers.** U.S. and Russian partners face a high risk of misunderstanding each other's intentions and of inadvertently creating discord in their relationships.

SOURCE: Office of Technology Assessment, 1995.

U.S. companies should proceed cautiously and develop clearly defined objectives for cooperative ventures.

FINDING 2: *Intergovernmental cooperation with Russia in space science, Earth observations, and space applications is developing well for the most part, although severe Russian budgetary constraints have put some projects in jeopardy. The political, technical, and administrative risks involved are somewhat higher than they are in NASA's traditional cooperative relationships, but U.S. program managers understand them and have planned accordingly.*

Although cooperation with Russia in space has varied widely in intensity, it has a three-decade history. The United States and the Soviet Union began sharing data from weather satellites in 1966.[13] U.S. and Russian space scientists have cooperated at some level of interaction since the late 1960s.

Over the years, the United States has made some useful gains in space science and Earth observations by cooperating with Russia. For example, data acquired during the Soviet Union's Venera Venus landings of the 1970s provided U.S.

---

[13] The first experimental Soviet weather observation sensors were flown in 1964. The sharing of weather-satellite data began two years later, after the launch of the Soviet Kosmos 122 satellite. (N.L. Johnson, Kaman Sciences Corporation, Colorado Springs, CO, personal communication, Feb. 6, 1995.)

BOX 1-3: Current and Future Intergovernmental Scientific and Technical Cooperation Between the United States and Russia

- **Solar system exploration:** proposed coordinated or joint missions to Mars, Pluto, and the neighborhood of the Sun; flight of instruments on each other's spacecraft, including Russia's Mars 96.
- **Space physics:** coordinated observing campaigns to study cosmic rays and the Earth's solar magnetic environment, using U.S., European, Japanese, and Russian spacecraft and ground facilities.
- **Space biomedicine, life support, and microgravity:** flight experiments on Mir, Spacelab, the Space Shuttle, and Russian biosatellites to support increased understanding of microgravity phenomena and factors affecting humans in space.
- **Earth sciences and environmental modeling:** flight of U.S. Earth Observing System instruments on Russian meteorological satellites; ground-based, aircraft, satellite, and spacecraft measurements of crustal and atmospheric phenomena and other aspects of the Earth as a system.
- **Astronomy and astrophysics:** flight of x-ray and gamma-ray instruments on each other's spacecraft; data exchanges and coordinated research.

SOURCE: Office of Technology Assessment, 1995.

---

planetary scientists with unique insights into chemical and physical processes in the Venusian atmosphere and on its surface.[14]

As noted above, throughout the Cold War, the overall state of the relationship between the United States and the Soviet Union placed general limits on the extent of cooperation in space activities. In addition, the risk that U.S. technology might be used to further Soviet military capabilities also limited the scope and depth of such cooperation from the U.S. side. Furthermore, Soviet authorities were loath to open their facilities to westerners or to allow their scientists and engineers to travel outside the Soviet Union. Nevertheless, during the Cold War, U.S. officials viewed scientific cooperation between the United States and the Soviet Union as helping to provide important insights into the workings of the closed Soviet society. Cooperation also exposed Soviet scientists to Western economic prosperity and political ideals. Since 1991, the pace of cooperative intergovernmental science and technology

programs with Russia has increased significantly (box 1-3).

Russia operates a fleet of reliable, relatively inexpensive launch vehicles. However, Russian space and earth sciences instrumentation and science spacecraft are generally not up to U.S. standards of sophistication and long-term reliability.[15] Russian strengths lie in theoretical science, materials, software development, space propulsion, and mechanical engineering. To counter their technical weaknesses, the Russians have actively sought foreign instruments for their spacecraft. Flying U.S. instruments on some Russian spacecraft continues to be an attractive way for the United States to gain additional flight opportunities at minimum cost. For example, from August 1991 until February 1995, a NASA Total Ozone Mapping Spectrometer (TOMS) delivered important data from aboard a Russian Meteor-3 polar-orbiting weather spacecraft. NASA has recently concluded an agreement with the Russian Space

---

[14] James W. Head III, "Scientific Interaction with the Soviet Union: The Brown Geology Experience," *Geological Sciences Newsletter*, May 1988, pp. 1-3.

[15] Russian engineers have compensated for less reliable spacecraft design by building operational spacecraft, such as their Meteor weather-monitoring satellites, in series and by launching new ones as needed. Russia's efficiency in launching payloads to Earth orbit makes this approach feasible.

Agency (RSA) to fly a Stratospheric Aerosol and Gas Experiment (SAGE) instrument and a TOMS instrument on Meteor-3M spacecraft in 1998 and 2000, respectively. At the same time, the U.S. participants in these cooperative efforts must be keenly aware of the risk that Russian agencies or enterprises may be late or unable to perform because of technical, economic, and/or political difficulties.[16] Well-developed contingency plans are, therefore, a necessity.

In the past, NASA has almost always arranged its cooperative projects so that there is no exchange of funds with the other countries or agencies involved. OTA's workshop participants believed that under normal circumstances, this practice was sound because it helps ensure balanced projects and avoids the political difficulties that could arise from sending funds abroad. Moreover, they agreed that foreign agencies that find a place in their budgets for their part of cooperative projects tend to be more fully engaged and committed partners.

Consistent with this approach, the use of Russian launch vehicles for U.S. space science and applications spacecraft is an attractive cooperative option that may permit some projects that would not be undertaken otherwise. Normally, such cooperative agreements should be made on the basis that Russia would supply the launch vehicle in return for participating in the activity and receiving access to the data returned. Nevertheless, as a short-term measure, U.S. support of some portion of the launch costs for an experiment may be appropriate—for example, to ensure the project's completion.[17]

Cooperation in space science and applications has a lower profile, and a less immediate connection to political matters, than does cooperation in human spaceflight and thus would be much more likely to survive a cooling in the political relationships between the United States and Russia. Nevertheless, the emergence of sharp policy differences between the two countries, particularly in geographical areas of intense conflict, might make all cooperative projects harder to carry out.

FINDING 3: *Because of the high cost, complexity, and public visibility of projects involving human spaceflight, cooperation on such projects promises the potential for both greater benefits and higher risks than it does in space science and applications. Although Russia's technical contributions to the International Space Station will result in a substantial increase in the station's planned capabilities, the potential benefit for the United States in this and other human-spaceflight projects lies at least as much in foreign policy as in space activities.*

Russia has more operational experience with long-duration human spaceflight than does the United States. During the 1970s and early 1980s, the Soviet space program orbited and operated six Salyut space stations. In February 1986, the U.S.S.R. launched the core module for the larger and more capable Mir Space Station, which is still

---

[16] For example, Russian budgetary constraints have forced the near abandonment of the "Mars Together" concept for linking the U.S. and Russian programs for the exploration of Mars in a series of joint missions.

[17] To obtain much-needed technical information about the Russian Meteor-3M spacecraft for determining the feasibility of joint missions, the United States has paid Russia about $100,000 for a set of spacecraft-interface design-and-control documents. These arrangements will help defray Russia's costs for its part of the Meteor-3M/SAGE and Meteor-3M/TOMS projects. The United States will also reportedly pay integration costs for the 1998 and 2000 flights, which will total around $5 million for both missions.

BOX 1-4: Technical and Operational Advantages of Cooperating with Russia on the International Space Station

- The use of Russian launch vehicles for construction and logistics, in addition to Space Shuttle and other Western vehicles, significantly improves transportation availability.
- The space station "Alpha" redesign with Russian participation will be completed 15 months earlier than the "Alpha" redesign without Russian participation (but two years later than Space Station Freedom's scheduled completion).
- The space station will have 25 percent more usable volume (if the European Columbus module is reauthorized in October 1995).
- When assembly is completed, the station will have 42.5 kilowatts more electrical power than did the "Alpha" design.
- Crew size can be increased from four to six, providing additional crew time for scientific experiments and maintenance.
- Portions of all international laboratory facilities will be within the zone of best microgravity conditions for research.
- An orbital inclination of 51.6° means that the space station will overfly a large portion of the Earth's surface, thus increasing opportunities for Earth observations.

SOURCES: "Addendum to Program Implementation Plan," NASA, Nov. 1, 1993; Marcia S. Smith, "Space Stations," Congressional Research Service Issue Brief 93017, Washington, DC, October 1994 (updated periodically); Office of Technology Assessment, 1995.

---

in operation under Russian control.[18] Throughout the program, cosmonauts have extended Russia's experience in long-duration spaceflight, up to the current record of 473 days.

In December 1993, U.S. Vice President Al Gore and Russian Premier Viktor Chernomyrdin announced their governments' agreement to cooperate on the International Space Station.[19] This agreement was a highly visible sign that the United States was willing to work more closely with Russia on important science and technology programs. It was undertaken in large part to underscore the new political relationship between the two countries, in which the United States and Russia are attempting to work together on technical and political problems of mutual interest. Also in December 1993, NASA Administrator Dan Goldin and RSA Director General Yuri Koptev signed a cooperative agreement for joint Space Shuttle-Mir experiments and a letter contract committing

the United States to a total of $400 million ($100 million per year for four years) for Russian goods and services related to the Shuttle-Mir program. The joint activities planned under this agreement should yield critical life sciences data and important insights into working with the Russians. Russia will make a substantial addition to the International Space Station by contributing several major components; the United States is purchasing other components (box 1-4). The United States will spend nearly $650 million in Russia over four years for the Shuttle-Mir program and other Russian space goods and services.

As noted earlier, by including the Russians in high-profile projects in space, U.S. officials hope to reduce possible proliferation of Russian military technology and assist the stabilization of Russian economic and political institutions. Successful execution of the space station agreement would also be an important symbol of the chang-

---

[18] Indeed, the two cosmonauts aboard Mir during the 1991 attempted coup were launched as citizens of the U.S.S.R. and returned to Earth as citizens of Russia.

[19] Al Gore and Viktor Chernomyrdin, Joint Statement, Dec. 16, 1993.

ing world order—a demonstration both of the ability of two former superpower adversaries to substitute cooperation for competition and of Russian integration into a major Western cooperative venture.

The Clinton Administration's policy of involving Russia in the International Space Station and other space projects also stems from a growing U.S. appreciation of Russian technical capabilities in developing and maintaining the support structure for humans in low Earth orbit (LEO). The series of engineering experiments beginning in 1995 involving Mir and the Space Shuttle will serve as important precursors to space station construction.

Involving the Russians in the International Space Station promises to increase program flexibility and capability. It also reduces the potential for space station failure resulting from the loss of a shuttle orbiter.[20] Russia has highly capable launch systems that can assist in building the space station and in supporting its operations and that can reduce the probability of interruptions in these activities. For example, Russia will contribute Soyuz-TM spacecraft for crew rotation and rescue (if needed) during the 1997-2002 period (box 1-5). However, including the Russians in the space station also increases the managerial complexity of space station planning, construction, and operations.

How the United States manages the relationship with its other space station partners and Russia will also affect space station success. The 1993 U.S. decision to invite Russia to participate in the International Space Station was the latest episode in a series of trials that have strained the partnership since the signing of the initial agreements in 1988. U.S. officials angered other partners by the unilateral manner in which they invited the Russians to join the project.[21] Since then, NASA has endeavored to repair the damage its actions might have done to an effective partnership and has taken care to involve fully *all* partners in space station decisions. Although working-level cooperative activities appear to be proceeding well in the new framework, other events may place added pressure on the space station partnership. In 1994, Canada decided to reduce its space station participation significantly, and the scale and shape of the European Space Agency's (ESA's) commitment remains uncertain, pending a ministerial meeting scheduled for October 1995.

FINDING 4: *Including Russia in the International Space Station program provides technical and political benefits to the space station partners, but placing the Russian contribution in the critical path to completion also poses unprecedented programmatic and political risks.*

Russian contributions to the space station involve the development and construction of several critical elements (box 1-6). To keep space station construction on schedule and costs down, these space station elements must be delivered on time and within budget. Because successful completion of the space station is so important to NASA's future, difficulties in meeting space station cost or schedule goals will especially stress NASA's relations with Congress and the other partners. Some analysts, for example, worry that although Russian supplies of space hardware seem adequate

---

[20] As noted in an earlier OTA report, the risk of losing a shuttle orbiter during or after space station construction is sufficiently high to raise concerns about the wisdom of using only the Space Shuttle to support the space station. The availability of Russian space-transportation systems greatly reduces the risk of a failure to complete space station construction. See U.S. Congress, Office of Technology Assessment, *Access to Space: The Future of U.S. Space Transportation Systems*, OTA-ISC-415 (Washington, DC: U.S. Government Printing Office, May 1990), p. 7.

[21] According to news reports, U.S. officials initially failed to consult adequately with Canada, the European Space Agency, and Japan concerning the inclusion of Russia in the joint project. See, e.g., "Clinton Orders New Design for Space Station," *Aviation Week and Space Technology*, Feb. 22, 1993, pp. 20-21; James R. Asker, "NASA's Space Station Takes Friendly Fire," *Aviation Week and Space Technology*, Mar. 22, 1993, p. 25; and "Station Partners Blast U.S. Design," *Aviation Week and Space Technology*, May 3, 1993, p. 20.

BOX 1-5: The SOYUZ-TM

The Soyuz series of crew ferries began with the Soyuz (1967-81), moved to the Soyuz-T (1979-86), and is represented today by the Soyuz-TM (1986-present). Designed and manufactured by NPO Energia (now RSC Energia), the Soyuz-TM carries a crew of two to three in its 10-cubic-meter habitable volume. Two 10.6-meter solar arrays provide electrical power during the two- to three-day trip from Earth to the Mir Space Station and connect to the space station's electrical system to provide it with an additional 1.3 kilowatts of electrical power. The Soyuz-TM is just over 7 meters long, has a gross weight of 7.07 metric tons, and is rated for flight times of up to 180 days while docked with Mir. The 20 successful launches of the Soyuz-TM, in addition to the 51 launches of the Soyuz and Soyuz-T, have given the Russians unparalleled experience in automated and manual docking procedures.[1]

The SOYUZ-TM Spacecraft

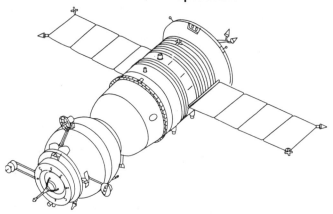

SOURCE: David F. Portree, *Mir Hardware Heritage*, 1994.

Part of the Russian contribution to the International Space Station will be Soyuz-TMs to serve for crew return and crew rotation during Phase Two of the space station. The Progress-M, a vehicle designed like the Soyuz-TM but made to carry cargo only, will be used throughout all phases to bring fuel and supplies to the space station.

[1]U.S. astronaut Norman Thagard and two Russian cosmonauts, Vladimir Dezhurov and Gennadiy Strekalov, were launched toward Mir aboard a Soyuz-TM on Mar. 15, 1995. On Mar. 16, they docked automatically with Mir. Astronaut Thagard will return to Earth on the Space Shuttle, which is scheduled to dock with Mir in June 1995.

SOURCE: N.L. Johnson and D.M. Rodvold, *Europe and Asia in Space, 1991-1992*, DC-TR-219/103-1 (Colorado Springs, CO: Kaman Sciences Corporation, 1993).

today, Russian enterprises might not be able to maintain an appropriate pace and quality of production.[22] On the other hand, for certain items, such as launch vehicles and many launch subsystems, Russian enterprises have excess capacity, and given sufficient funding, may be able to increase their production to meet market needs.

At least one OTA workshop participant observed that Russian space enterprises have functioned extremely well in building a production series of spacecraft but have had difficulties meeting schedules with new, untried, and one-of-a-kind designs. Others expressed concern about the state of second- and third-tier equipment suppliers

---

[22] Judyth L. Twigg, "The Russian Space Program: What Lies Ahead?" *Space Policy* 10(1): 19-31, 1994.

BOX 1-6: Key Russian Elements in the Critical Path of the International Space Station

- Guidance, navigation, and control of the Phase Two station depends on the Functional Cargo Block (FGB) module, purchased from Khrunichev Enterprise by Lockheed.
- Reboosting the space station, to prevent premature reentry caused by atmospheric drag, will depend on a series of Russian Progress-M and Progress-X cargo spacecraft (the latter is an enlarged version that does not yet exist).
- Russia will be responsible for crew-return ("lifeboat") capability in the period before planned space station completion (1997-2002).
- Fuel resupply is also a Russian-only function, under current plans.

SOURCE: Marcia S. Smith, "Space Stations," Congressional Research Service Issue Brief 93017, Washington, DC, October 1994 (updated periodically).

---

to the large enterprises and about whether those suppliers could continue to meet planned production schedules.

Other concerns relate to Russia's space infrastructure. For example, the Baikonur Cosmodrome, or launch facility, in the nation of Kazakhstan is a crucial part of Russia's contribution to the space station program; the Proton, Soyuz, and Zenit launch vehicles are launched from Baikonur. Visitors to the facility in 1994 expressed concern about its condition and about low morale among key personnel at the site.[23] Further, news reports about Leninsk, the city created to support the launch complex, have painted a grim picture of living conditions.[24] However, prospects for the long-term viability of the Baikonur Cosmodrome have improved as a result of: 1) the ratification of the Russia-Kazakhstan agreement on its use and 2) the apparent resolution of internal Russian government differences over funding its operation and maintenance. Recent visitors from NASA and Anser Corporation report that the Baikonur facilities are in good repair. In addition, LKE International's decision to invest in the modification of a payload processing facility at Baikonur demonstrate that at least one major American-Russian partnership has confidence in the long-term operation of the Cosmodrome. Finally, in 1994, Russia

launched 13 Proton vehicles from Baikonur, which tied the all-time high for the number of Proton launches in a given year.

Of perhaps greatest importance to the relationship between Russia and the United States in the International Space Station effort is whether political and/or military events within Russia or between Russia and other countries will cause either the United States or Russia to amend or even cancel their space station agreement. Although few believe that a resumption of the level of hostility present during the Cold War is likely in this restructured world, a sharp cooling of the U.S.-Russian political relationship could slow or even cancel space station activity. On the other hand, the space station's high visibility in the overall political relationship and the Clinton Administration's strong commitment to the project could help insulate it from transitory political strains.

FINDING 5: *Although the emerging commercial partnerships between the United States and Russia exhibit promise in some space sectors, it is too early to tell how successful these partnerships will be. Because of the higher economic and political uncertainties, commercial ventures with Russian companies carry much higher risk than those with firms in Western Europe or Japan.*

---

[23] *Oversight Visit: Baikonur Cosmodrome*, Chairman's Report of the Committee on Science, Space, and Technology, House of Representatives, 103d Congress, 2d Session (Washington, DC: U.S. Government Printing Office, February 1994).

[24] "Baykonur-Leninsk Difficulties Evaluated," Foreign Broadcast Information Service, FBIS-USR-94-074, July 5, 1995.

BOX 1-7: Areas of Commercial Cooperative Activities

- **Propulsion technology.** Using Russian engines to enhance the performance of existing U.S. launch vehicles, or for potential use in future systems, such as reusable demonstration vehicles.
- **Launch services.** Using Russian space-launch vehicles for Western satellites.
- **Launch-vehicle components.** Taking advantage of Russian expertise in materials and fabrication to achieve cost and weight savings and increased reliability.
- **Telecommunications services.** Using Russian communication satellites to provide international services.
- **Others.** Remotely sensed data; underlying technologies and materials; software and analytical services.

SOURCE: Office of Technology Assessment, 1995.

Russian companies have much to offer U.S. companies, especially in liquid-fuel propulsion, launch vehicles, and launch services. Incorporation of Russian technologies into U.S. launch vehicles and launch operations could make U.S. launch services more competitive in the international marketplace than they are today. Russian launchers such as Proton and Soyuz are highly capable and have a strong record of launch success. Russia also has significant skills in satellite remote sensing and now markets, through several Western companies, the highest-resolution remotely sensed data[25] that are commercially available.[26] However, Russian skills in the global marketplace are just developing. The combination of Russian know-how and U.S. marketing skills can improve the international competitiveness of U.S. companies and also help Russian companies earn much-needed hard currency.

Continued progress in developing these partnerships (box 1-7) will depend in part on the speed with which Russia converts its large state enterprises into firms driven by market forces. It will also depend on the development of stable Russian laws and institutions aimed at reducing the institutional uncertainties and economic risks of such ventures. Commercial progress will also depend on developing and maintaining stable and supportive U.S. and Russian governmental policies toward these industrial partnerships.

As noted above, the Clinton Administration, with cautious encouragement from Congress, has pursued policies of greater openness with the FSU. U.S. agencies could play a significant role in easing the path of private sector cooperative agreements with Russia, but interagency conflicts and continued distrust of Russian motives impede greater progress. Despite the end of the Cold War, the dismantlement of the Coordinating Committee on Export Controls (COCOM), and the realignment of the State Department and Commerce Department's export-control responsibilities (during which many space items were removed from the State Department's U.S. Munitions List), U.S. export-control restrictions continue to increase the time and cost involved in cooperating with Russian entities in both the public and private sectors.

Although business relationships among U.S. and Russian firms have generally developed without unduly affecting either side's relations with third parties, competition in launch services may

---

[25] However, the highest-resolution Russian remotely sensed data are in photographic form and cannot be provided in a timely manner after acquisition, which inhibits their use in applications where timeliness or well-calibrated radiances are necessary.

[26] WorldMap International, Ltd., markets data from the Russian Resurs-F satellites that have been digitized from photographic originals; several other companies market photographic images or digital data from second-generation images obtained through Soyuzcarta, a Russian data-marketing firm.

prove an exception. The U.S. government faces growing pressure from some U.S. firms and Russia either to liberalize its launch-services agreement with Russia or abandon it altogether.[27] The latter, in particular, seems likely to provoke a strong protest from Europe, which favors an upper limit on launches for Russia.

The political and economic uncertainties in Russia should prompt U.S. companies to be cautious in pursuing partnerships with newly created Russian private companies. In addition, the changing institutional relationships within the Russian government make navigating Russian regulatory requirements a challenge to U.S. companies. Russian officials worry about the loss of economically or militarily significant technologies to the West. For most space technologies, the Russian military plays an important role in the establishment of fruitful business relationships with Russian companies. Despite the increasing power of Russian aerospace corporations to chart their own destiny, many U.S.-Russian agreements require the consent and/or the active participation of the Russian Ministry of Defense and the Military Space Forces. These parties, which may possess veto power over projects, are often not included in negotiations at the early stages.

FINDING 6: *The Russian government has made important strides in reorganizing its civilian space program to allow smoother cooperation with Western governments and commercial enterprises. Nevertheless, Russia's space program faces many challenges in achieving long-term stability.*

To change the way space policy is made within Russia, to separate the civilian space effort from the military, and to make cooperation with other governments and with non-Russian corporations easier, the Russian government created the Russian Space Agency in February 1992. Made up of a relatively tiny cadre of about 200 people who were part of the old Ministry of General Machine Building that once controlled all aspects of the space program, RSA reports directly to the government of Russia. It is responsible for drafting space policy and for implementing the policy once it has been ratified by the government. Funding for RSA comes through several ministries, including the Ministry of Science. RSA is responsible for space program management, budgeting, and international negotiations. The agency lacks the personnel to engage in R&D activities or detailed program oversight, and it must depend on Russian industry to carry out many of the functions that NASA's field centers perform in the United States.

Whatever the prognosis for the commercial space industry, Russian space science will likely suffer during the next few years. The Institute of Space Research (IKI) is the official body that orchestrates Russia's efforts in space science, and although IKI is a part of the Russian Academy of Sciences (RAS), it depends on the RSA for its funding. Funding for science will almost certainly take a back seat to funding for projects that are either necessary to the state or that promise to bring in Western currency. Two other organizations are involved in determining Russia's space science efforts: 1) the Interdepartmental Scientific and Technical Council on Space Research (MNTS KI), which provides peer review of proposed projects and is chaired by the head of RAS, and 2) the Interdepartmental Expert Commission, which is made up of chief designers from industry and members of other ministries and which tries to coordinate the needs of industry with those of the scientific community.

FINDING 7: *The Russian space program is suffering from the current political and economic climate in the FSU. The budget for space activities is decreasing sharply. The survival of some parts of Russia's space program will depend on cooperation with other countries.*

---

[27] The agreement, which was signed in September 1993, limits Russian commercial launches to eight between 1993 and 2000. See chapter 5, "Governments as Regulators."

The Russian space-related work force has decreased 30 to 35 percent over the past three years. A total of 200,000 workers have left the industry for more lucrative aerospace jobs elsewhere, both inside and outside Russia. Many of these workers are young people who are leaving for more promising futures in the emerging private sector. One of the largest aerospace firms, RSC Energia, used to hire 2,000 young people each year; now it hires 200. From December 1994 through February 1995, RSA argued forcefully in the Russian media and before the Duma that unless its 1995 funding request was met, the space program could collapse. Although the Duma was publicly sympathetic, there is no reason to expect that RSA will fare better in 1995 than it did in 1994, when it reportedly received approximately one-quarter of the funding it requested, and one-half of what had been allocated by the government. On the other hand, the space program has already survived what may be the worst times. Efforts at restructuring the space program and moving to a market-oriented way of doing business have ameliorated the situation and are continuing. Systems required by the government (such as reconnaissance satellites) will continue to be funded, but funds for basic research, new designs, and many commercial projects will almost certainly have to come from external sources. Despite many economic hardships, the Russian civil and military space programs continue to lead the world in annual numbers of launches and active satellites.

FINDING 8: *NASA's purchase of goods and services from Russia serves important foreign policy goals and improves the chances that Russia will be able to meet its obligations to the International Space Station. The scale of NASA funding that this requires, however, further increases the political risk faced by the International Space Station program.*

The planned payment of nearly $650 million from the NASA budget during FY 1994-97 (directly and through contractors) for Russian space goods and services represents by far the largest transfer of funds from the U.S. budget to Russian government and private organizations in that period.[28]

Symbolically, these payments are an important international signal of U.S. support for Russia's transition to a market economy. Given Russia's pride in its aerospace accomplishments, U.S. support for that sector takes on added political and psychological significance. U.S. purchases of Russian space goods and services should also help to sustain Russian adherence to the Missile Technology Control Regime, both because of the political linkage that was established when the $400 million NASA/RSA contract was announced[29] and because the funding will help preserve employment for Russian engineers and technicians in at least some major Russian space-industrial centers. By funding a significant portion of the total RSA budget and making payments to Russian enterprises that play pivotal roles in the Russian contribution to the International Space Station program (such as RSC Energia and the Khrunichev Enterprise), NASA improves the chances that Russia will be able to meet its obligations to the space station project.

On the other hand, the size of the funding requirement virtually guarantees that it will be controversial when considered by Congress, particularly in the context of efforts to reduce the U.S. budget deficit. Moreover, some observers have questioned the wisdom of supporting the Russian aerospace industry, which provided much of the technological underpinnings for the Soviet threat to the United States. However, most, if not all, of these funds would be spent in industrial subsec-

---

[28] See chapter 3, "The Financial Dimension," for a detailed discussion of these payments.

[29] The White House, "Joint Statement on Space Cooperation," from the first meeting of the U.S. Russian Joint Commission on Economic and Technological Cooperation, Washington, DC, Sept. 2, 1993.

tors that support spaceflight rather than ballistic-missile production.

FINDING 9: *The French experience in cooperating with the Soviet Union and Russia since 1966 largely parallels and confirms that of the United States.*

France and the European Space Agency have the two most significant programs of space cooperation with Russia other than the United States. The long French relationship with the FSU demonstrated an early understanding and acceptance of the importance of political motives for space cooperation.[30] French President DeGaulle's 1966 decision to begin cooperating with the Soviet Union on space projects was principally intended as an assertion of French independence within the Western alliance, but it quickly acquired significant substantive content, particularly in the space sciences. In 1982, France and the Soviet Union began a series of cooperative human-spaceflight activities, despite strains in the political relationships with Western nations caused by Soviet actions in Poland. These actions precipitated a U.S. decision, in the same year, to allow formal space ties to lapse. In contrast, the French opted to maintain cooperative ties, adjusting the scale of cooperation in response to the state of the political environment.

ESA, which is a relatively new participant in cooperating with Russia, is now spending significant amounts on several major projects. In all, ESA committed about $81 million to Russia to pay for Mir flights and other activities between November 1992 and the end of 1994. ESA has budgeted approximately $240 million to provide European-built hardware for use on Russian elements of the space station.

FINDING 10: *Although the Russian government and Russian enterprises have preserved most of the technical and managerial capabilities of the former Soviet Union, Ukraine also retains significant space assets and capabilities. Kazakhstan owns the Baikonur launch facility and several tracking and data stations. The United States may find it beneficial to form partnerships with firms and governmental entities in these countries.*

- *Kazakhstan.* Russia and Kazakhstan have concluded a long-term agreement on the support and use of the Baikonur Cosmodrome, which is a critical component in Russia's launch infrastructure. The importance of this launch complex to the launch of space station components and supplies guarantees U.S. interest in the continuation of good political relations between Russia and Kazakhstan. NASA is exploring cooperation in environmental research, space science, and telemedicine with Kazakhstan and is maintaining its own lines of communication with Kazakhstani space authorities in order to follow space developments there closely.

- *Ukraine.* Russia itself uses launch vehicles with significant Ukrainian content[31] and Ukrainian-built components extensively.[32] The United States will have to determine the appropriate balance between working directly with Ukrainian partners and developing ties through Russia. So far, the U.S. approach has been to rely on Russia to represent Ukraine in matters in-

---

[30] See appendix D and U.S. Congress, Office of Technology Assessment, *U.S.-Soviet Cooperation in Space*, OTA-TM-STI-27 (Washington, DC: U.S. Government Printing Office, July 1985), ch. 4.

[31] An example is the Zenit booster, for which NPO Yuzhnoye, Ukraine, is the prime contractor and for which the Russian firm NPO Energomash provides the main engines.

[32] However, some Russian enterprises are cutting back their dependence on Ukranian suppliers of space goods in a government-wide effort to make the Russian space program independent of Ukraine. See Peter B. deSelding, "Russia Distances Space Program from Ukraine," *Space News*, Feb. 20-26, 1995, p. 3.

volving the space station partnership, while awaiting confirmation of Ukrainian adherence to the Missile Technology Control Regime (MTCR).[33] Meanwhile, NASA is seeking to advance cooperation with Ukraine in areas such as environmental research and telemedicine.

FINDING 11: *Despite the economic and political uncertainties, most early participants in cooperative ventures have found the potential gains worth the problems of pursuing cooperative projects. Participants suggest that any organization planning cooperation with Russian (or other former Soviet) organizations take several precautions to enhance the success and minimize the risks of such projects.*

- *Plan for the possibility of nonperformance.* Given the significance of the Russian contribution to the space station, the U.S. ability to make up for delays or for failure to deliver is severely limited by available U.S. resources.[34] In robotic space exploration, program managers emphasize the importance of such planning from the outset.
- *Seek a better understanding of the larger political and economic forces that could affect Russia's ability to deliver on commitments.* Some increased confidence might be obtained through further systematic analysis of Russian adaptation of their defense industry to post-Soviet conditions.
- *Maximize open and frank communication.* Minimizing technical and managerial surprises means seeking (and allowing) a high degree of communication and interpenetration between U.S. programs and their Russian counterparts, both for the space station partnership and in robotic space cooperation.
- *Be prepared for delays and reverses, and seek good advice.* Businesspeople interviewed by OTA believe that the best protection against the

immaturity of Russia's legal and business systems is to obtain sound advice from Russian experts, to expect delays and reverses, and to be patient.

- *Be aware of and manage cultural differences effectively.* As noted in finding 1, cultural differences can also increase the level of project risk. To minimize these risks, U.S. entities should:

  ▸ Sensitize all personnel who will be in contact with Russian personnel to be aware of cultural differences, learn ways to avoid affront, and build personal rapport with their Russian counterparts.
  ▸ Resist the temptation to assume that U.S. and Russian personnel share common assumptions about the meaning of business or contractual terms and concepts; when in doubt, such terms should be spelled out. Find out who has the authority to make the needed decisions.
  ▸ Avoid postures or assumptions of superiority, particularly in technical areas; a good rapport and mutual respect for each other's technical achievements and capabilities are critically important.
  ▸ Make use of the best available expertise in Russian nonaerospace business law and practices, both to structure relationships properly and to avoid surprises as much as possible when political or financial circumstances change.

FINDING 12: *Experts disagree over the extent to which cooperation with the Russian government and industry on space projects would affect the retention of U.S. jobs.*

Some industry officials have expressed concern that U.S. jobs could be lost as a result of using Russian technology in the U.S. space program. Others have argued that skillful incorporation of

---

[33] Ukraine has agreed to abide by the restrictions of the MTCR, but before being admitted to the regime, it must demonstrate its adherence to the terms of the regime.

[34] The United States cannot afford to maintain parallel developments for the Russian FGB module or the Soyuz-TM crew-return vehicle.

Russian technologies in U.S. projects could save taxpayer dollars in publicly funded programs such as the space station and could boost U.S. international competitiveness in commercial programs. Although the use of Russian technologies and know-how may cause some job shifts, and even the loss of certain technical skills, if U.S.-Russian cooperative activities are properly structured, they could improve the scope of the U.S. space program and, possibly, enhance U.S. competitiveness.

Russian launch vehicles and related systems (particularly propulsion systems) have the most obvious potential for commercial use. Russian launch experience is unmatched, and both existing hardware and underlying technological developments can fill important gaps in U.S. capabilities. On the other hand, U.S. national security interests demand that the United States maintain its national launch capability and technology base. The simple purchase of vehicles or launch services appears to be less attractive than joint ventures, co-production of vehicles and/or systems, and analogous business arrangements as ways of accommodating these differing interests.

History and Current Status of the Russian Space Program | 2

I n 1957, the Soviet Union put the first satellite, Sputnik, into orbit. In 1961, it launched the first human, Yurii Gagarin, into space. From that time until its dissolution in 1991, the U.S.S.R. maintained a robust space program, often following lines of development very different from its one major competitor, the United States. However fast the political and economic landscape may be changing in Russia, the speed with which the space program can change and the directions it can take are constrained by how it developed during the Soviet era. This chapter gives a synopsis of the legacy of the Soviet space program.[1] It then describes what we know about the current status and structure of the Russian space program and what direction it might take in the next few years.

THE SOVIET SPACE PROGRAM

Russia's civilian space program is still using equipment and material manufactured and stored before the dissolution of the U.S.S.R., such as stockpiles of Proton rockets, satellites, and the Mir Space Station. Some of the impetus for the high-level production was a desire to equal or surpass U.S. accomplishments in space. Figure 2-1, which shows the number of launches in the United States and in the U.S.S.R. since 1957, not only demonstrates the productive capacity of the U.S.S.R.'s space industry, but also indicates the difference in design philosophies of the two countries. Where the United States built long-lived, technically

[1] Much of the material in this chapter is drawn from N.L. Johnson and D.M. Rodvold, *Europe and Asia in Space, 1991-1992*, DC-TR-2191.103-1 (Colorado Springs, CO: Kaman Sciences Corporation, 1993).

FIGURE 2-1: Successful U.S. and U.S.S.R./Russian Launches

SOURCE: Marcia S. Smith, *Space Activities of the United States, CIS, and Other Launching Countries/Organizations: 1957-1993*, Congressional Research Service Issue Briefs, Washington, DC, Mar. 29, 1994.

sophisticated payloads, the U.S.S.R. built much shorter-lived satellites that required more frequent replacement.

The difference in design philosophy between the two countries goes back to the origins of their space programs. At the end of World War II, the German rocket scientists from Peenemünde, who were responsible for the V-2 Rocket, were part of the spoils of war divided between the United States and the U.S.S.R. Both countries used the experience and skills of these men to set up their ballistic-missile programs. Because the Soviet hydrogen bomb was so much larger and heavier than the one developed in the United States, it required a larger, more powerful rocket to carry it. In fact, the Soviet Union produced the first intercontinental ballistic missile (ICBM) in the world, which was known in the West as the R-7. The Soviets' expertise in producing rockets with large lift capacities then made it possible for them to be the first to produce launchers that could carry humans into orbit.

The successful production of rockets with large lift capacities reduced incentives to make payloads compact and light. To this day, the Russian satellite is less sophisticated in its electronics (Soviet satellites continued to use vacuum tubes long after the West had switched to solid-state components) and heavier than its Western counterparts. The Soviets built satellites with a much shorter design life than was typical in the United States. The requirement to maintain these space systems led to the Soviet Union's remarkable (by Western standards) ability to replace damaged or obsolete satellites. For some types of satellite, the Soviet Union was able to launch a replacement in 24 to 48 hours.

Both in the United States and the Soviet Union, the space program was a symbol of the country's technological superiority and productive capacity. The United States kept its military program out of the public eye and created the National Aeronautics and Space Administration (NASA) as a separate civilian space program, with its own budget, as the focus of the national civilian space effort. The Soviet Union, on the other hand, never created separate civil and military space programs; the same budget supported both efforts. Much of the same infrastructure, production organizations, design bureaus, and personnel were used to service both programs. Furthermore, there was no functioning legislative body that determined the budget; instead, funds went directly from the government to the design bureaus and production organizations via the ministries (figure 2-2). Ultimately, the Central Committee of the Communist Party of the Soviet Union decided what was to be funded. Which projects the Central Committee considered worthy of funding depended in part on what the United States was doing at the time and who among the industrial and military leaders had the government's ear.

The procurement of any system for the space program, civil or military, began with an order from the Council of Ministers. Money was then appropriated for the Ministry of General Machine Building (MOM), which passed it directly to the plant or design bureau chosen to do the work. The same funds would be used to do the systems tests, and if the tests went well, the Military Industrial Commission (VPK) would place the order for large-scale production. Once all the pieces were in place to produce a system that fulfilled requirements and had passed testing, the system's rigidity would deter attempts to infuse innovative and untested technological changes. Often, one well-tested design for a spacecraft would be used in widely disparate parts of the space program. Commonly, a spacecraft designed for the Soviet human space program would be used for robotic purposes. For example, the Soviet Vostok spacecraft, the type that carried Gagarin into orbit, was modified to become the Zenit photographic-reconnaissance spacecraft and, also, the Photon materials-processing platform.

THE RUSSIAN SPACE PROGRAM

▪ The Breakup of the U.S.S.R.

In December of 1991, the Union of Soviet Socialist Republics was dissolved. Of the 15 republics of the U.S.S.R., the three Slavic states (Russia, Ukraine, and Belarus) joined the Commonwealth of Independent States (CIS) on December 8; the five central Asian states (Kazakhstan, Kyrgyzstan, Tajikistan, Turkmenistan, and Uzbekistan) and the three transcaucasian states (Armenia, Azerbaijan, and Moldava) joined on December 28; and Georgia joined in 1993. The Baltic states (Latvia, Lithuania, and Estonia) are not members of the CIS. At the end of December 1991, to keep the former Soviet space program intact, Russia led an attempt to form a CIS space agency.[2] That organization has turned out not to be influential, and although Ukraine, Kazakhstan, Azerbaijan, and Uzbekistan have all formed their own space agencies, Russia is by far the dominant player in the post-Soviet space program.

With the dissolution of the U.S.S.R. has come the daunting task of establishing a new form of government in Russia and, after decades of economic and political isolation from the West, altering a command economy to make it competitive with Western markets. The Politburo and Secretariat of the Communist Party are no longer the chief decisionmakers in Russia. A newly instated[3] Duma, or legislative assembly, controls the appropriations process, while the executive ministries are subordinate to the prime minister, who is

---

[2] The "Minsk Space Agreement," signed December 30, 1991, by representatives of the republics of Azerbaijan, Armenia, Belarus, Kazakhstan, Kyrgyzstan, Tajikistan, Turkmenistan, Uzbekistan, and the Russian Federation, lays out general guidelines for continuing the U.S.S.R.'s space program through combined use of resources and proportionate funding. Ukraine signed in the summer of 1992. The document itself is kept in the archives of the Belarus Republic in the capital city of Minsk.

[3] On September 21, 1993, Russian President Boris Yeltsin dissolved the Supreme Soviet of the Russian Federation. Two weeks later, on October 3 and 4, he used the Russian Army to suppress resistance to his order by a group of deputies from the dissolved Soviet. Yeltsin ordered a referendum on a new constitution and elections on December 12, 1993. The resulting election created a bicameral legislature, which had as its upper house the Federation Council and as its lower house the historically named State Duma. For an account of the conflict and the elections, see J. Nichol, Congressional Research Service, *Russia's Violent Showdown: Chronology of Events, September 21-October 4, 1993*, 93-879 F (Washington, DC: October 1993), and J. Nichol, Congressional Research Service, *Russian Legislative Elections and Constitutional Referendum: Outcome and Implications for U.S. Interests*, 94-19 FAN (Washington, DC: January 1993).

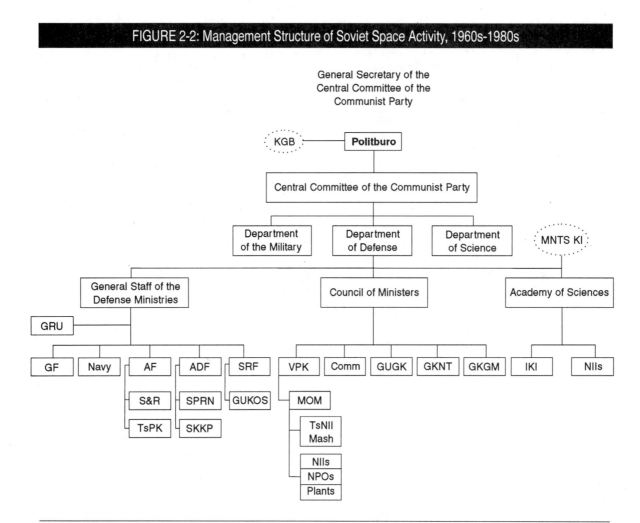

FIGURE 2-2: Management Structure of Soviet Space Activity, 1960s-1980s

KEY: ADF=Air Defense Forces; AF=Air Force; GF=Ground Forces; GKGM=State Committee for Geodesy and Mapping; GKNT=State Committee for Science and Technology; GRU=State Intelligence Directorate (Military Intelligence); GUGK=State Directorate for Hydrometrology; GUKOS=State Directorate for Space Means; IKI=Institute of Space Research; MNTS KI=Interdepartmental Scientific and Technical Council on Space Research; MOM=Ministry of General Machine Building; NII=Scientific Research Institute; NPO=Scientific Production Organization; S&R=Search and Rescue; SKKP=System for Monitoring Outer Space; SPRN=System for Warning of Missile Attack; SRF=Space Rocket Forces; TsPK=Gagarin Cosmonaut Training Center; VPK=Military Industrial Commission.

SOURCE: Maxim Tarasenko, Moscow Institute of Physics and Technology, 1994.

appointed by the president. As a consequence, the Russian space program faces difficulties in organizing a network of facilities that stretches across several independent countries. As long as the non-Russian republics were part of the U.S.S.R., facilities such as design bureaus, factories, a launch facility (cosmodrome), and many sites used for satellite telemetry and tracking were available to the ruling body in Moscow for its use. Now, the independent republics are in a position to demand payment for facilities on their land, for hardware, and for services.

Very few of the facilities outside Russia can be considered indispensable to the survival of its space program. However, the Baikonur Cosmodrome is one facility that Russia would have a

---

Box 2-1: Space Facilities and Organizations Outside Russia

Launch facility

The only launch facility in the former Soviet Union now outside Russia is the Baikonur Cosmodrome, in Tyuratam, Kazakhstan.

Organization for launch-vehicle and satellite design and manufacture

NPO Yuzhnoye (formerly the Yangel Design Bureau) is located in Dniepropetrovsk, Ukraine. Yuzhnoye produces the SS-18 and SS-24 ballistic missiles and the SL-11 (Tsyklon-M), SL-14 (Tsyklon), and SL-16 (Zenit) launchers. It also produces remote-sensing, intelligence, and weather satellites. At one time, NPO Yuzhnoye had nearly 30,000 employees.

Spacecraft command-and-control centers

The Space Command, Control, and Tracking System (KIK) has its main control centers near Moscow: 1) the Flight Control Center (TsUP) at Kaliningrad, which handles planetary missions, the Mir Space Station, and Soyuz missions to Mir, and 2) the Satellite Control Center (TsUS) at Golitsino, which handles all civilian and military satellites. KIK also has sites outside Russia in Georgia, Kazakhstan, Ukraine, and Uzbekistan. It controls the nearly 180 currently active Russian and Commonwealth of Independent States satellites. A subset of these sites is also used as the primary support network for the Mir Space Station.

The Long-Range Space Communications System (TsDKC), which controls scientific spacecraft in high Earth orbit or in interplanetary flight, has sites in Russia and Ukraine.

Space-surveillance facilities

The System for Monitoring Outer Space (SKKP)[1] and the System for Warning of Missile Attacks (SPRN) use HEN HOUSE and Large Phased Array Radar developed in the 1960s and 1980s. Sites outside Russia are in the Ukraine, Kazakhstan, Azerbaijan, and Latvia. SKKP also uses seven optical sensors located in Russia, Kazakhstan, Tajikistan, and Ukraine, and it uses five electro-optical sensors located in Russia, Armenia, Georgia, Turkmenia, and Ukraine.

---

[1] This facility is now referred to officially as the Space Surveillance System.

SOURCE: Kaman Sciences Corporation, 1994.

---

hard time replacing quickly.[4] Rather than build a new launch complex, the Russian government has decided that it is more effective and cheaper, for the time being, to lease the cosmodrome from Kazakhstan for the next 20 years at least.[5] Most of the spacecraft command, tracking, and control stations outside Russia have been taken offline and are being replaced with space-based autorelay satellites.[6] Russia must now deal with the production organizations and design bureaus that lie outside its borders as it would with any other foreign enterprise. Box 2-1 shows some of the facilities and organizations that now lie outside Russia.

---

[4] Although Baikonur is the usual name for the cosmodrome located near the Tyuratam railway station in Kazakhstan, the town of Baikonur lies some 320 km (200 miles) northeast of the cosmodrome.

[5] The Military Space Forces (VKS) and the Russian Space Agency (RSA) are struggling over whether or not to establish a new cosmodrome at the old ballistic-missile site in Svobodny. The VKS wants the security of launching all military payloads from Russian soil, but the RSA does not want its funding diluted by having money diverted away from Baikonur.

[6] The command and tracking stations in Ukraine are still online.

After drawn-out negotiations dating back to 1992, Prime Minister Viktor Chernomyrdin of Russia and his Kazakh counterpart, Akezhan Kazhegeldin, signed a treaty on December 10, 1994, that leases to Russia the use and control of Baikonur until the year 2014, with the possibility for a 10-year extension of the lease. In the near future, those same parties are expected to sign an agreement leasing to Russia several military test ranges that lie in Kazakh territory. One plan under consideration would convert the ballistic-missile launch site at Svobodny into a space-launch facility to reduce or eliminate dependence on Baikonur.[7]

■ The Russian Space Agency

With the dissolution of the U.S.S.R., a new governmental structure and, two years later, a new constitution were established in Russia. Russia's legislative structures started to play a more significant role in governing, and the executive branch revised its structure. (See figure 2-3 for the organization of the revised government as it bears on the space program.) Because of the importance of the space program to Russia's defense and economic well-being and because the space industry would have to compete in the world market to survive, the Russian government empowered a separate agency to control the direction of the state space program and to act as the representative of the Russian space program in dealings with the Newly Independent States (NIS) and other foreign countries.

The Russian Space Agency (RSA) was established by decree of the president of Russia on February 25, 1992, and was given its charter[8] by the legislative branch on October 6, 1993, with a mandate to "make efficient use of Russia's space-rocket complex in the interests of the Russian Federation's socio-economic development, security and international cooperation. . . ."[9] RSA is supposed to be funded as a separate item in the Russian federal budget, as it was in 1993 and is expected to be in 1995. In 1994, the funding for RSA was not a separate line item but came through the Ministry of Science. RSA's organizational structure, which is similar to NASA's, is shown in figure 2-4.[10] Under the Soviet system, there had never been any agency whose sole purpose was to formulate and implement government space policy. The existence of such an agency has changed the way the space program is run in several ways:

- *It is now possible to separate the civil and military parts of Russia's space program.* Though the military still commands a large portion of resources in the space program, Russia is moving away from having Space Forces personnel operate all launch and operations facilities. Within the next two years, RSA will be responsible for paying civilian personnel to take over functions that have to do with civilian launches and the maintenance and operation of civilian satellites and of the Mir Space Station. At present, RSA pays the salaries of 16,000 servicemen, who, although they are no longer paid by the Space Forces, are still under the command of the Commander of the Space Forces.

- *Russian facilities are now more accessible to the West because: 1) the military part of the program has been separated out, which makes transfer of sensitive military technology easier to control, and 2) fewer government departments have to sign off on a cooperative venture.* Furthermore, RSA provides a single point of contact for any organization wanting to do

---

[7] "New Launch Base Sought," *Aviation Week & Space Technology*, p. 54, Jan. 2, 1995.

[8] Law of the Russian Federation on Space Activity, Section II, Article 6, October 1993.

[9] Boris Yeltsin, Decree Establishing the Russian Space Agency, Moscow, February 1992.

[10] However, RSA is an organization of only 200 people, compared with NASA's workforce of 22,000. Lynn Cline, Director, Space Flight Division, Office of External Relations, NASA, points out that it is impossible for RSA to oversee implementation of cooperative agreements and contract awards in the way that NASA does, simply because of its small size.

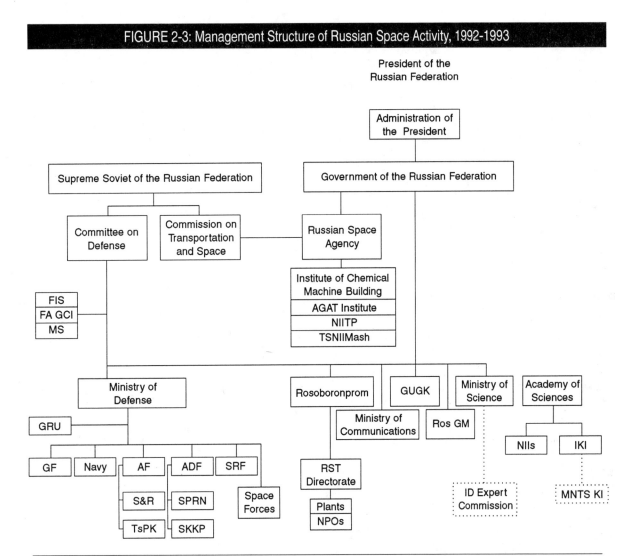

FIGURE 2-3: Management Structure of Russian Space Activity, 1992-1993

KEY: FA GCI=Federal Agency of Government Communication and Infcrmation; FIS=Foreign Intelligence Service; ID=InterDepartmental; MS=Ministry of Security; Ros GM=Russian State Committee of Hydrometrology; RST=Rocket and Space Technology. See figure 2-2 for definitions of other abbreviations.

SOURCE: *Who's Who in Russian Aerospace Activities: A Reference List*, ANSER, Center for International Aerospace Cooperation, December 1994.

business with Russia in commercial space endeavors, if the organization chooses to use it.[11] In that regard, RSA replaces Glavkosmos, which now primarily markets Russian technology. RSA's director, Yuri Koptev, must approve all contracts between foreign entities and the various enterprises of the Russian space program and is reportedly very willing to approve contracts that benefit those enterprises.[12]

---

[11] There are company-to-company contacts, but RSA involves itself in the negotiations at some level.

[12] Two Russian contacts, one a scientist with Applied Mechanics NPO and a member of the Russian Academy of Sciences, the other an analyst of the Russian space program, both agree that RSA is committed to helping the Russian aerospace industry do as much business with foreign corporations as possible.

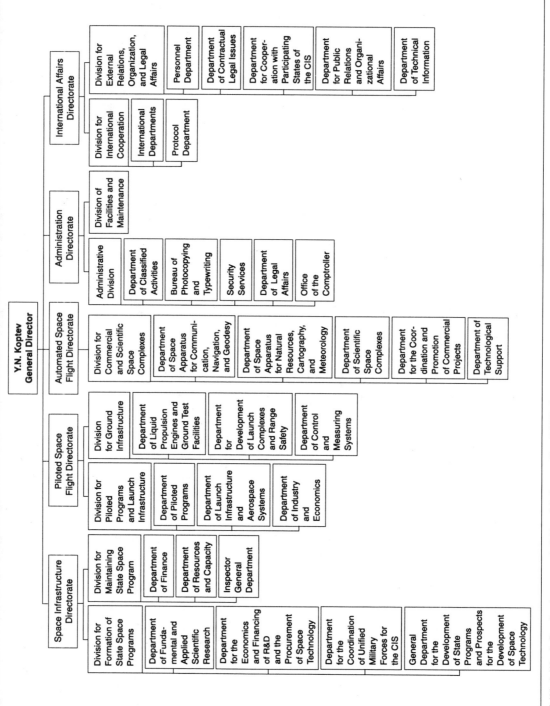

FIGURE 2-4: Russian Space Agency

SOURCE: *Who's Who in Russian Aerospace Activities: A Reference List*, ANSER, Center for International Aerospace Cooperation, Arlington, VA, December 1994.

- *RSA is involved in all phases of the development of a system, from research and development through production.* At its inception, the agency was given authority over several research and production organizations (figure 2-3), including the Central Scientific Research Institute of Machine Building (TsNIIMash), which operates the spacecraft control facilities of the Flight Control Center (TsUP) at Kaliningrad. After a struggle between RSA and several of the larger space enterprises, the Russian government, in a decree signed by Prime Minister Chernomyrdin,[13] gave control of 38 aerospace enterprises to RSA, to be added to the four companies[14] already under RSA control. This action shifted authority for the funding and oversight of those enterprises from the State Committee on Defense Industries to RSA. The change also means that RSA is now responsible for all defense conversion efforts.

RSA could also help make the transition to private enterprise less abrupt. In the recent privatization of one of Russia's largest and most influential space enterprises, Scientific Production Organization (NPO) (now Russian Space Corporation (RSC)) Energia, ownership of 38 percent of the company was retained by the government, that is, by RSA. Of the remaining 62 percent, 10 percent (120,000 shares at 1,000 rubles each) was offered to the employees, 5 percent was given to management, 10 percent was held in reserve, 12 percent was sold at auction, and 25 percent was exchanged for privatization vouchers. As Russian aerospace companies become more established in the world market, the extent to which RSA has to subsidize the infrastructure that supports those enterprises could decrease significantly.

■ Current Activities

The percentage of the total 1993 Russian Federation space budget devoted to the different areas of Russia's space program is shown in figure 2-5. These resources are used to support Russia's civilian and military objectives, namely,

- exploring space,
- pursuing space science,
- maintaining human presence in space via
  - ► space station and
  - ► cargo and logistics vehicles,
- maintaining information space systems, such as
  - ► navigation satellites,
  - ► geodetic satellites,
  - ► telecommunication satellites and
  - ► observation satellites, and
- maintaining space assets dedicated to national security.

Of these, the severe budgetary constraints of the past few years have curtailed efforts in space science and exploration the most.[15]

Human Spaceflight

Russia has had a frequent human presence in space since the Salyut program began in 1971. The Salyut Space Stations passed through several different designs before reaching the most recent phase, which is represented by the Mir Space Station. Mir has been in orbit since 1986 and has been permanently occupied since 1989. Although not all of the operations have gone according to plan,

---

[13] Government of the Russian Federation Decree #866, Moscow, July 25, 1994. The decree states, "The RSA will provide state control and coordination of enterprises and organizations involved in the research, design, and production of rocket and space hardware for various purposes; to determine state scientific, technical, and industrial policy in the areas of rocket and space hardware and to ensure the fulfillment of such policies; and to enable the fulfillment of conversion projects and the structural reorganization of the rocket and space industry."

[14] The original four enterprises under RSA control are the Scientific Research Institute of Chemical Machine Building (NIIKhimMash), the AGAT Institute, the Scientific Research Institute of Thermal Processes (NIITP), and the Central Scientific Research Institute of Machine Building (TsNIIMash).

[15] The U.S.-Russian mission to Mars, which was scheduled to fly in 1994, will be lucky to get off the ground as early as 1996. The "Fire and Ice" mission to explore Mercury and Pluto is now highly tentative. It stands to reason that in tight budgetary times, programs deemed less essential will be cut back first, and those programs needed for national security or that can attract outside revenues will be given priority.

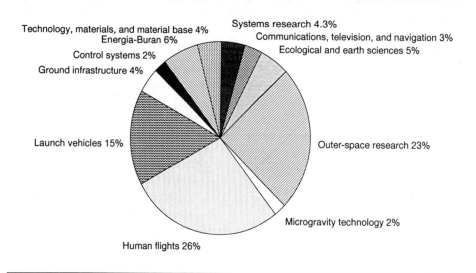

FIGURE 2-5: Distribution of the 1993 Russian Space Budget

Technology, materials, and material base 4%
Energia-Buran 6%
Control systems 2%
Ground infrastructure 4%
Launch vehicles 15%
Human flights 26%

Systems research 4.3%
Communications, television, and navigation 3%
Ecological and earth sciences 5%
Outer-space research 23%
Microgravity technology 2%

SOURCE: Kaman Sciences Corporation, 1995.

with several failures of the automatic docking system and occasional problems with power maintenance and environmental control, the Russians have used the Mir station to gain more experience in the adaptation of humans to the space environment than has any other nation.

The Mir Space Station has been used to perform international experiments in science and space engineering. It has been host to astronauts from the United States, England, Germany, Japan, and other countries. The use of the station has given the Russians unequaled experience in the effects of prolonged spaceflight, extravehicular activities, docking, and maintenance of facilities in space. Experiments on Mir include research in the fields of botany, biology, materials science, and physiology, many of which included international participation. Because of their experience with space stations, the United States expects Russia to play a large part in the design and maintenance of the International Space Station.

Civil Space Systems

Russia has approximately 180 operational satellites in over 30 active satellite programs; in addition, many inactive, standby satellites can be brought into use if needed. Box 2-2 summarizes the active systems that are not strictly military.

National Security Space Systems

An early impetus for the development of satellites was to use space for military observation, which is one reason Sputnik created such a scare.[16] Later came the realization that orbiting platforms could be used for early-warning systems and space weapons. The Russian antisatellite systems (ASAT) program dates back to 1963, and the first ASAT intercept test was performed in 1968.

Russia currently operates several types of military-reconnaissance satellite[17]:

- Low-Earth-orbit (LEO) high-resolution satellites that fly for up to three months with 1-meter

---

[16] Even before the time of Sputnik, both the United States and the Soviet Union understood the surveillance and communications potential of satellites.

[17] N.L. Johnson, Kaman Sciences Corporation, Colorado Springs, CO, personal communication, February 1995.

(or better) resolution. The photographic film returns to Earth and is picked up.

- Satellites with both photographic film return and digital transmission capabilities. These satellites remain in orbit for up to two months.
- Satellites with lifetimes of a year or more that transmit their digital data to Earth via relay from geosynchronous-orbit (GEO) satellites.
- Topographic mappers that fly for six weeks with 2-meter-panchromatic and 5-meter-color resolutions.
- Vostok-class satellites which fly for two to three weeks and are assumed to function like the Resurs-F remote-sensing satellites (box 2-2).
- A new satellite, designated Kosmos 2290 and launched on the Zenit, which is still under assessment.
- EORSAT (Electronic Ocean Reconnaissance), a four-satellite constellation that flies at altitudes of about 400 km. The system is believed to be able to estimate naval positions to within 2 km.
- Russia flies two other types of electronic intelligence satellite. One, launched on the Tsyklon, flies at an altitude of 650 km; the other, launched on the Zenit, flies at 850 km altitude.

Two constellations support ballistic-missile-attack warning systems, and one supports an ASAT system. One of the early-warning systems has nine satellites in Molniya orbits equipped with infrared sensors (see box 2-2 for an explanation of the Molniya-type orbit); the other has two or three satellites in GEO. The ASAT system operates by waiting for the launch site at Baikonur to pass through the orbital plane of the offending satellite, at which time the ASAT would be launched on an intercept path requiring one or two revolutions, or between 90 and 200 minutes. A conventional explosive device would then destroy the target satellite.

■ Launch Systems

Russia maintains a versatile fleet of launch vehicles capable of lifting payloads of half a metric ton to 21 metric tons into LEO. All Russian- and Ukrainian-built boosters, their primary launch sites, principal manufacturers, and payload capacities are listed in the table in figure 2-6. These vehicles are launched from the two cosmodromes currently in use, Plesetsk in Russia and Baikonur in Kazakhstan.

Baikonur Cosmodrome

Baikonur is the oldest space-launch facility in the world and has supported more than 968 orbital missions since 1957.[18] Information from visiting westerners and from wire and news reports confirm that the infrastructure is deteriorating, that many buildings where work is going on are unheated, and that certain parts of the complex are in disrepair.[19] There were reports of strikes and protests last winter because of the harshness of conditions and the lack of basic amenities in the neighboring town of Leninsk. Reportedly, bands of thieves were even using camels to pull copper cable from the ground, melting the cable down, and selling the copper for its value as raw material. The Russian press reported that one launch at Baikonur had to be postponed because of the theft of specialized equipment.[20]

The picture is not altogether bleak, though. At the cosmodrome itself, maintenance and modifications are being kept up to allow for the continued launching of all families of booster traditionally launched out of Baikonur. Under the

---

[18] About 70 missions failed to reach Earth orbit.

[19] See, e.g., *Oversight Visit: Baikonur Cosmodrome*, Chairman's Report of the Committee on Science, Space, and Technology, House of Representatives, 103d Congress, 2d Session (Washington, DC: U.S. Government Printing Office, February 1994), Margaret Shapiro, "Site of First Steps into Space Slips into Poverty," *The Washington Post*, Mar. 19, 1995, p. A1, and Michael Specter, "Where Sputnik Once Soared into History, Hard Times Take Hold," *The New York Times*, Mar. 21, 1995, p. C1.

[20] Vladimir Ardayev, "Kazakhstan: Depressing Landscape With Fireworks," *Isvestiya*, Moscow, Nov. 3, 1994, p. 4 (Foreign Broadcast Information Service translated text).

Box 2-2: Operational Russian Satellite Systems

Telecommunications satellites

- Molniya ("Lighting") is a constellation of 16 satellites in highly elliptical (eccentricities[1] of about 0.75), semisynchronous orbits used for telephone, telegraph, and television transmission. The Molniya-type orbit was designed by the U.S.S.R., and the satellites view broad regions of the Northern Hemisphere for eight hours out of their 12-hour periods. The Molniya-type orbit has the advantage of giving better coverage of the Earth at high latitudes than does a geostationary orbit. By choosing an orbit with an inclination of 63.4°, the satellite's point of closest approach remains fixed in the Southern Hemisphere, thereby ensuring that the satellite always flies over the same region in the Northern Hemisphere.
- Ekran ("Screen") satellites are geostationary at 99° east longitude and provide television and radio transmission for Russia's far-northern regions. At present, two such satellites are operational.
- Gorizont ("Horizon") is a high-power telephone, telegraph, television, radio, and fax transmission satellite, which also handles maritime and international communications. Through Gorizont, Russia has become increasingly more integrated into the Western system of communication satellites. When the U.S.S.R. became a member of INTELSAT in 1991, it made Gorizont available to that system, and several Western nations now lease Gorizont transponders. The United States leased transponders during the Persian Gulf War to handle the increased communications traffic in the Middle East.
- Gals is a television broadcasting system made up of one satellite in geosynchronous orbit (GEO). Originally meant as a part of a national program, the satellite is currently serving foreign customers.
- Luch is a three-satellite GEO system used for teleconferencing, television, video data exchange, and telephone links. To date, only two of the three GEO satellites have ever been operational, and the Luch has been used primarily in support of the Mir Space Station.
- Raduga and Geyser (Geyser carries the Potok data-relay system) are constellations of satellites used for military and government communications.

Remote-sensing satellites

- Resurs-F1, first flown in 1979 with a nominal lifetime of 14 days, performs low-altitude (250 to 400 km) multispectral photography with cameras of 5- to 8-meter and 15- to 30-meter resolutions, and returns the film.
- Resurs-F2, first flown in 1987 with a nominal lifetime of 30 days, flies at altitudes of 170 to 450 km and returns multispectral photographs with 5- to 8-meter resolution.
- Almaz-1, first flown in 1991 (after a 1987 prototype), has a nominal lifetime of two years, flies at altitudes of 250 to 350 km, and uses a 15- to 30-meter-resolution synthetic aperture radar.
- Resurs-O, first flown in 1985 (with prototypes during 1977-1983), has a nominal lifetime of three years and performs multispectral imagery in Sun-synchronous orbit at altitudes of 600 to 660 km.
- Okean-O, first flown in 1983, with a nominal lifetime of two years, performs multispectral sensing and real aperture radar measurements for oceanographic surveys at altitudes of 630 to 660 km. This satellite can look at ice in the polar regions and spot weak points as an aid to navigation. Okean-O has also proved useful in search-and-rescue operations.

agreement between the Russian and Kazakh governments whereby Russia will lease the cosmodrome, Russia gains control over all of Baikonur's facilities and will pay the Kazakh government the equivalent of $115 million per year. Some of that money can be in the form of services and support of Leninsk, which is home to the workers who are still employed at Baikonur.

Facilities currently in use at Baikonur are reported to be in good working order and operating

---

Box 2-2 (Cont'd): Operational Russian Satellite Systems

Meteorological satellites

- Meteor-2, first flown in 1975, carries a scanning photometer, an infrared radiometer, and a radiation measurement complex. It is now being phased out. Meteor-3, first flown in 1984, carries a direct-scanning telephotometer, a store-and-dump scanning telephotometer, a direct-scanning radiometer, a store-and-dump scanning radiometer, a UV spectrometer, a multichannel UV spectrometer (Ozon-M), and a radiation measurement complex. In 1991, the United States and Russia cooperated on a project to place NASA's Total Ozone Mapping Spectrometer (TOMS) on a Meteor-3 spacecraft.

- Electro, once called the Geostationary Operational Meteorological Satellite (GOMS), has been under development since the 1970s. Launched October 31, 1994, it carries a scanning infrared radiometer, a scanning telephotometer, and a radiation measurement complex.

Material processing and biological satellites

- A Photon spacecraft, dedicated to materials science research, has flown every year since 1988, except for 1993. International organizations also make use of the Photon for their microgravity experiments. Bion, like the Photon, uses a Vostok-like recoverable spacecraft to perform life sciences experiments and to return the payload to Earth. Ten Bion flights have occurred over the past 20 years, and the last seven have all had extensive international, including U.S., participation.

---

[1] The eccentricity of an orbit is a measure of how far that orbit is from being circular. (A circular orbit has an eccentricity of zero.) Satellites have their greatest velocity at perigee and their smallest velocity at apogee. The Molniya orbit is designed to have its perigee in the Southern Hemisphere which puts the apogee in the Northern Hemisphere where the satellite needs to spend most of its time to be most effective.

SOURCE: Kaman Science Corporation, 1994.

---

with the automated efficiency typical of the Russian launch industry. Baikonur supported 23 missions in 1993; and 30 in 1994. In addition, the U.S. firm of Lockheed Aerospace is building a payload-processing facility in support of its joint-venture subsidiary Lockheed-Khrunichev-Energia, and construction has begun on special buildings dedicated to the processing of Western payloads.

Russia launches five boosters from Baikonur —Rokot, Tsyklon-M, Soyuz, Zenit, and Proton (three-stage and four-stage). Energia, with and without the shuttle Buran, is no longer operational, and the Molniya booster has not flown since 1989. The Baikonur Cosmodrome supports Russian programs in human spaceflight, interplanetary spaceflight, communications and early-warning satellites in GEO, navigation and geodetic satellites, remote-sensing satellites, satellites used for national security purposes, and scientific satellites (including interplanetary satellites).

Plesetsk Cosmodrome

The cosmodrome at Plesetsk was, until last year, the world's busiest space-launch facility. It has averaged one launch per week for the past 10 years, and it has launched nearly 1,400 missions since 1966, including 26 missions in 1993 and 18 in 1994. Plesetsk is capable of supporting launches of Start-1, Kosmos, Tsyklon, Molniya, and Soyuz. Plesetsk supports remote-sensing, meteorological, communications, navigation, and scientific satellites, as well as satellites used for national security purposes.

Some Russian officials would like to see all boosters launched from Russian soil. One plan would upgrade Plesetsk so that it could launch

FIGURE 2-6: Russian and Ukrainian Boosters

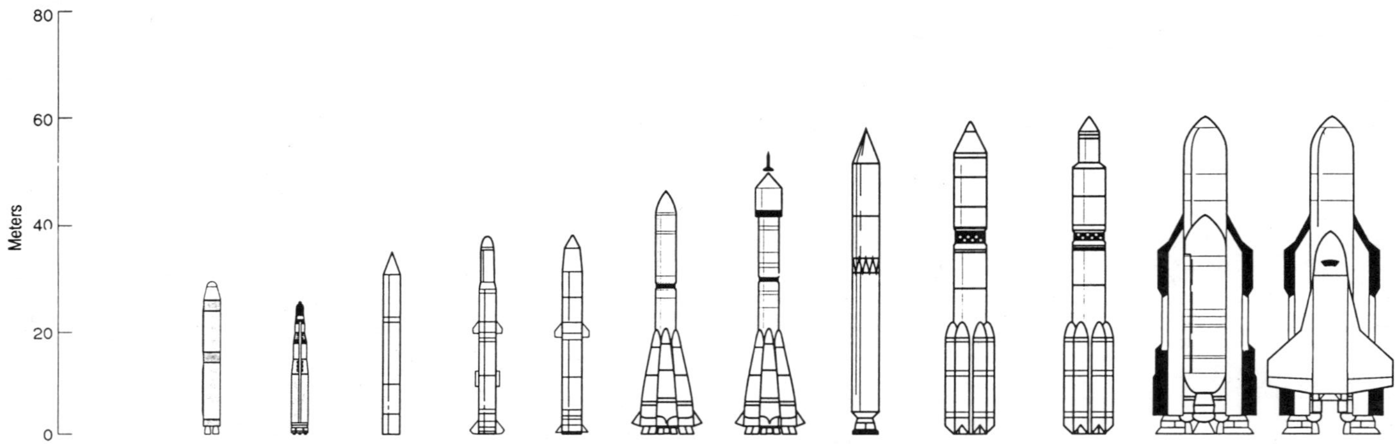

Designator	SL-19	SL-18	SL-8	SL-11	SL-14	SL-6	SL-4	SL-16	SL-12	SL-13	SL-17	SL-17
Name	Rokot	Start-1	Kosmos	Tsyklon-M	Tsyklon	Molniya	Soyuz	Zenit	Proton	Proton	Energia*	Energia/Buran*
First Flight	1991	1993	1964	1966	1977	1961	1963	1985	1967	1968	1987	1988
Current Launch Sites	Baikonur	Plesetsk	Plesetsk	Baikonur	Plesetsk	Baikonur Plesetsk	Baikonur Plesetsk	Baikonur	Baikonur	Baikonur	Baikonur	Baikonur
Principal Manufacturer	Salyut Design Bureau	Yuzhnoye NPO	Yuzhnoye NPO	Yuzhnoye NPO	Yuzhnoye NPO	TsSKB Samara	TsSKB Samara	Yuzhnoye NPO	Khrunichev Enterprise	Khrunichev Enterprise	Energia NPO	Energia NPO
Configuration	3-Stage	4-Stage	2-Stage	2-Stage	3-Stage	3-Stage	2-Stage	2-Stage	4-Stage	3-Stage	1-Stage	1-Stage
Stage 1	SS-19 Stage 1	SS-25 Stage 1	RD-216	RD-218	RD-218	RD-107,108	RD-107,108	RD-170	RD-253	RD-253	RD-0120/171	RD-0120/171
Stage 2	SS-19 Stage 2	SS-25 Stage 1	N/A	RD-219	RD-219	RD-0110	RD-0110	RD-120	RD-0210,0211	RD-0210,0211		
Stage 3	Breeze	SS-25 Stage 1			N/A	N/A			RD-0212	RD-0212		
Stage 4	N/A								Block D/DM			
Payload (tonnes)	2.0 (LEO)	0.5 (LEO)	1.5 (LEO)	3.5 (LEO)	4.0 (LEO)	1.8 (SEO)	7.32 (LEO)	13.7 (LEO)	2.5 (GEO)	20.6 (LEO)	88.0 (LEO)	30.0 (LEO)
Uses	N/A	N/A	Early-warning mission	Military payloads, e.g., ASAT, EORSAT, RORSAT	Meteor, Okean, geodetic	Molniya, Kosmos	All manned; Photon, Bion, Resurs-F, Progress-M	Military ELINT	Solar System exploration, communications, navigation, early warning	Space Station components	N/A	Buran Space Shuttle
Precursor	SS-19	SS-25	SS-5 (Skean)	SS-9 (Scarp)	SS-9 (Scarp)	SS-6 (Sapwood)	SS-6 (Sapwood)	None	None	None	None	None

Meters: 80 — 60 — 40 — 20 — 0

*The Energia and Energia/Buran are not currently operational.

SOURCE: Kaman Sciences Corporation, 1995.

TABLE 2-1: Possible Russian Ballistic-Missile Conversion

Name	Ballistic missile	Payload to LEO[1] (metric tons)	Launch site
SS-18K	SS-18	4.4	Baikonur
Space Clipper	SS-24	2.	Air launch near Equator
Shtil	SS-N-23	0.95	Air launch near Equator
Reef	SS-N-20	1.5	Air launch
Surf	SS-N-20/SS-N-23	2.4	Sea launch near Equator

[1]Low Earth orbit.

SOURCE: Kaman Sciences Corporation, 1994.

every vehicle in the fleet, but the financial constraints make that kind of construction unlikely in the near term. A second proposal is to convert the missile range at Svobodny into a cosmodrome for civilian and military launches. The Russian Duma has not approved the installation of facilities to launch the proposed Angara heavy-lift launch vehicle, but it has not prohibited the planned 1996 operations of the Rokot small-payload launcher. Engineers have already begun converting the old ICBM silos for launching the Rokot, and the first test launch from Svobodny is planned for summer or fall of this year.

∎ Military Conversion

The end of the Cold War leaves Russia with several classes of ballistic missile that could be converted to commercial lift vehicles. Besides the Start-1 and Rokot, which are already operational, Russia will test several conversions in the next few years. Conversion is costly, and the number of systems that become operational will depend on market demand. Table 2-1 shows the ballistic missiles that Russia has considered for conversion to commercial launchers, along with their payload capacities and launch sites.

The Cooperative
Experience
to Date $\big|$ 3

PUBLIC SECTOR

Cooperation on civil space projects with the world's other space superpower has been discussed and sometimes pursued since the beginning of the Space Age, although during the Soviet period, competition generally dominated.[1] Before 1991, the ability to pursue cooperation was frequently compromised by the vicissitudes of the Cold War because the linkage between space cooperation and broader superpower relations frequently worked to restrict even modest projects. For example, the United States allowed the government-to-government agreement on the cooperative use of space to lapse in 1982 over Soviet imposition of martial law in Poland.

Although linkage to political concerns continues, it currently works to stimulate rather than limit cooperative activity. Moreover, with serious space-budget shortfalls across the rest of the spacefaring world, most observers of the U.S. space program consider extensive international cooperation, involving Russia as well as traditional partners, essential to the achievement of national goals in space. This section briefly traces the history of public sector space cooperation between the United States and the

---

[1] For a detailed review of international cooperation and competition up to 1985, see U.S. Congress, Office of Technology Assessment, *International Cooperation and Competition in Civilian Space Activities*, ISC-239 (Washington, DC: U.S. Government Printing Office, June 1985). See also U.S. Congress, Office of Technology Assessment, *U.S.-Soviet Cooperation in Space*, TMI-STI-27 (Washington, DC: U.S. Government Printing Office, July 1985). The standard political history of this period in science and technology, with particular attention to space cooperation and competition between the United States and the Soviet Union, is Walter A. McDougall, . . . *The Heavens and the Earth: A Political History of the Space Age* (New York: Basic Books, Inc., 1985).

Soviet Union (and later, its successor states) and describes its status through early 1995.

The Early Years: 1958-1971

Even before the launch of Sputnik 1, the United States sought to engage the Soviet Union in space cooperation on two broad fronts—diplomatically, through proposals to guarantee the peaceful use of outer space, and scientifically, through the machinery of the International Geophysical Year (IGY).[2] Both countries explicitly linked their initial satellite efforts to the IGY. After Sputnik 1, both the Eisenhower Administration and Congress gave heightened emphasis to calls for scientific collaboration.[3] Relatively little tangible

cooperation resulted, however, because the competitive element predominated on both sides.

Even before his inauguration, President John F. Kennedy commissioned an extensive study of potential space cooperation with the Soviet Union and signaled this interest both in his Inaugural Address and in his first State of the Union message, as part of a broader effort to engage the U.S.S.R. in cooperation in relatively nonsensitive areas. The study arrived at the White House on April 14, 1961, two days after Yurii Gagarin's first orbital flight.

The space-cooperation study contained more than 20 individual proposals, ranging from arms-length scientific collaboration to proposals for establishing a joint lunar base. U.S. prestige around the world suffered dramatically because of Gagarin's flight, and as a result, the balance of U.S. attention shifted to competition, particularly after President Kennedy's announcement of the Apollo Program on May 25, 1961. However, a first, modest agreement on space cooperation between Moscow and Washington was reached in 1962; it provided for a limited exchange of weather-satellite data, coordinated satellite measurements of the Earth's magnetic field, and communications experiments involving the U.S. Echo II satellite. Results were mixed, and cooperation in satellite meteorology, in particular, was slow to begin.

Civil Space Agreements, Apollo-Soyuz, and Shuttle-Salyut: 1971-1982

The race to the Moon ended in 1969. Meanwhile, in 1967, the United States and the Soviet Union reached a political accommodation in the United Nations (U.N.) Outer Space Committee, resulting in the U.N. Treaty on Principles Governing the Activities of States in the Exploration and Use of Outer Space, Including the Moon and Other Celestial Bodies.

BOX 3-1: The Apollo-Soyuz Test Project

- The objective of ASTP was to develop and demonstrate compatible rendezvous and docking systems for U.S. and Soviet manned spacecraft. The docking mechanism to be used during the seven-flight Shuttle-Mir program is an improved variant on the ASTP design.
- On July 17, 1975, three U.S. astronauts and two Soviet cosmonauts docked Soyuz 19 with an Apollo spacecraft that was carrying the jointly developed docking module. Soyuz 19 and Apollo undocked after two days of symbolic visits between spacecraft.
- ASTP was widely praised as a symbol of détente, while also criticized at the time as an expensive symbolic gesture that was wasting scarce U.S. space funds.
- Follow-on Shuttle-Salyut mission preparations were suspended in 1978 amid worsening U.S.-U.S.S.R. relations.

SOURCE: Office of Technology Assessment, 1995.

---

[2] The IGY was established in 1957 by the International Council of Scientific Unions to pool international efforts in studying the Earth, the oceans, the atmosphere, and outer space.

[3] For a detailed discussion of cooperation before 1974, see Dodd L. Harvey and Linda C. Ciccoritti, *U.S.-Soviet Cooperation in Space* (Miami, FL: University of Miami, Center for Advanced International Studies, 1974).

Early in the 1970s, the general political thaw between the United States and the U.S.S.R. extended to space cooperation. A series of senior-level meetings between the National Aeronautics and Space Administration (NASA) and U.S.S.R. Academy of Sciences delegations in 1970-71 resulted in agreements on the organization of the Apollo-Soyuz Test Project (ASTP) and on cooperation in satellite meteorology; meteorological sounding rockets; research on the natural environment; robotic exploration of near-Earth space, the Moon, and the planets; and space biology and medicine. The 1972 Agreement on Cooperation in the Peaceful Exploration and Use of Outer Space, signed at the summit by Presidents Richard M. Nixon and Alexei Kosygin, formalized these understandings and endorsed the Joint Working Group (JWG) structure that had emerged to implement ASTP and to develop specific cooperative projects (see box 3-1 and photo above).

Work on ASTP proceeded relatively smoothly, although both sides approached the flight with suspicion and caution. Meanwhile, modest but mutually satisfactory cooperation—largely restricted to exchanges of data and coordinated experiments of various types—was developing in the areas of space science and applications, particularly in space biology and medicine.

Not long after the successful ASTP flight in 1975 (figure 3-1), the two countries agreed to pursue a follow-on rendezvous and docking activity involving the U.S. Space Shuttle (which had not yet flown) and the Soviet Salyut Space Station (figure 3-2). Shuttle-Salyut was the centerpiece of the renewal of the intergovernmental agreement between the U.S.S.R. and the United States in 1977 under President Jimmy Carter, which otherwise extended the 1972 agreement's provisions. Although extensive science planning for Shuttle-Salyut was completed in 1978, U.S. enthusiasm for the venture began to wane as relations cooled because of conflicts over human rights in the U.S.S.R. and, later, Soviet international actions. Concern about the possible technology-transfer implications of ASTP led to an extended interagency review, which found the program innocent

NASA Administrator James Fletcher, with Apollo 16 astronauts, briefs President Richard Nixon on the Apollo-Soyuz Test Project.

of any technology losses, though it acknowledged that the Soviets had probably learned a good deal about NASA's management of large projects. The study also recommended a careful, arms-length approach to additional cooperation, with structured interagency review of all proposals.

In 1978 and 1979, U.S. (and perhaps Soviet) interest in Shuttle-Salyut diminished further. The White House decided not to schedule the next technical meeting, which the United States had agreed to host. In 1979, President Carter mandated a sharp reduction in remaining activity under the 1977 agreement, following the Russian intervention in Afghanistan. In late 1981, with the imposition of martial law in Poland, the Reagan Administration announced that in retaliation, the civil space agreement would be allowed to lapse in May 1982.

■ Hiatus and Improvisation: 1982-1987

In the absence of an agreement, U.S. officials authorized only low-profile cooperation, with approval on a case-by-case basis by the White House. Despite this stricture, a certain amount of activity continued. COSPAS-SARSAT, a satellite-aided search-and-rescue project involving cooperation between the SARSAT partners (the United States, Canada, and France) and the Soviet COSPAS program, was judged by the White House to have overriding humanitarian value and

FIGURE 3-1: Apollo and Soyuz Join in Space

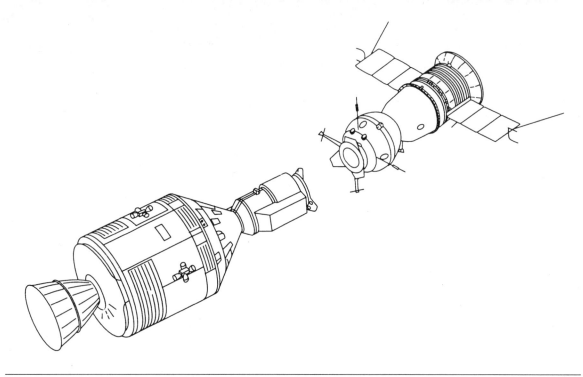

SOURCE: David S.F. Portree, *Mir Hardware Heritage*, Houston, TX, 1994.

operated uninterrupted.[4] NASA was allowed to continue to pursue cooperation in space biology and medicine, which, along with planetary data exchanges, had produced the most valuable scientific results under the 1972 and 1977 agreements; U.S. biomedical instrumentation flew on Soviet biosatellite missions in 1983 and 1985. Planetary-data exchanges also continued, principally involving studies of the atmosphere and surface of Venus.

In 1981, the space agencies of the United States, the U.S.S.R., Europe, and Japan formed the Inter-Agency Consultative Group (IACG) for Halley's Comet, an informal coordinating framework for the upcoming Halley's Comet apparition. Both the United States and the Soviet Union were members of the IACG, and NASA's Deep Space Network provided most of the tracking support for the European Space Agency's (ESA's) Giotto and for the Venus and Halley encounters of the U.S.S.R.'s VEGA-1 and 2. U.S. scientists also participated in data exchanges and joint analyses with Soviet counterparts through the IACG. In addition, several U.S. or partly U.S. instruments actually flew on the Soviet spacecraft, by virtue of

---

[4] SARSAT stands for Search and Rescue Satellite-Aided Tracking, and COSPAS is from the Russian for "Space System for the Search of Vessels in Distress."

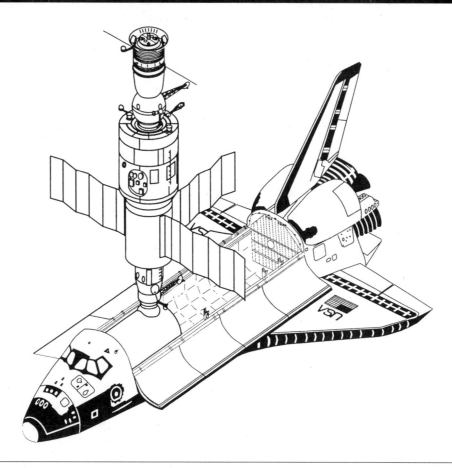

FIGURE 3-2: Conceptual Drawing of the Shuttle Docked with Salyut

SOURCE: David S.F. Portree, *Mir Hardware Heritage,* Houston, TX, 1994.

agreements negotiated with third parties who, in turn, concluded agreements with the U.S.S.R.[5]

As the Reagan Administration began to feel its way toward an improved relationship with the U.S.S.R., the first tentative steps were taken toward resumption of more formal, high-profile space cooperation. In January 1984, days before President Ronald Reagan's State of the Union address, in which he invited U.S. friends and allies to participate in the construction of a space station, the U.S. privately proposed to the Russians the idea of a simulated space-rescue mission involving the U.S. Space Shuttle and the Salyut-7 Space Station. Publicly and privately, the Russians were cool to the idea, perhaps because of the perceived asymmetry of a mission in which the Space Shuttle would simulate a rescue of cosmonauts from Mir. That summer and for the next two years, the U.S.S.R. also insisted on a linkage between progress in space arms control and a willingness

---

[5] In one instance, a U.S.-built flight instrument for the Vega mission was actually subjected to formal interagency review and approved for export to Russia on the ground that it was "not sophisticated" enough to be considered space hardware. A second instrument for the Phobos missions to Mars was on its way through a similar review process in December 1984, when the builder of the first instrument publicly proclaimed that he had outmaneuvered the Washington bureaucracy, angering the reviewing agencies and foreclosing further approvals at that time.

to consider expanded civil space cooperation, effectively precluding forward movement on the latter.

In mid-1986, however, the situation changed dramatically. In an exchange of letters between General Secretary Mikhail Gorbachev and President Reagan, Gorbachev dropped the arms-control requirement. Moscow accepted a U.S. proposal for an exploratory meeting in Moscow in September, at which U.S. and Russian delegations discussed and agreed upon a 16-item list of areas for expanded cooperation. The agreement text itself was negotiated in Washington at the end of October 1986, and in April 1987, rather than wait for a summit meeting, the two sides signed the agreement at the foreign minister level.

■ *Glasnost* and the End of the Soviet Era: 1987-1991

The 1987 agreement, which owed much of its restrictive structure and provisions to the 1970s experience, differed importantly from its predecessors by including an annex with a list of 16 approved areas for cooperation. It resurrected the JWG structure and authorized the formation of groups in space biology and medicine, solar system exploration, astronomy and astrophysics, space physics, and earth sciences. The JWGs were expected to meet at least annually. Amendments to the annex, announced at a succeeding summit in May 1988, authorized the exchange of instruments for flight on robotic spacecraft, as well as the exchange of planning data on future missions. Interagency approval was not forthcoming, however, for activity in human spaceflight going beyond research in space medicine or for higher-profile robotic cooperation in Mars exploration.

In August 1991, the United States and the U.S.S.R. achieved an important milestone with the flight of the U.S. Total Ozone Mapping Spectrometer (TOMS) on a Soviet Meteor-3 polar-orbiting meteorological satellite. More than two years elapsed between the agreement in principle and the conclusion of the U.S.-U.S.S.R. Memorandum of Understanding on the flight, a delay largely attributable to intensive U.S. interagency negotiations on technology-transfer controls. Finally, against the background of the political evolution in Eastern Europe and Russia, and given the importance of continuity in the collection of atmospheric ozone data, a compromise was reached. Shortly after the successful launch, while U.S. engineers and scientists were still in Moscow for checkout activities, the abortive anti-Gorbachev coup was launched, signaling the beginning of the end for the Soviet Union.

■ Current Cooperation in Space Science and Applications

The U.S.S.R.'s collapse and the emergence of separate Russian, Kazakhstani, and Ukrainian states dramatically changed the political context for space cooperation. The linkage between political interests and cooperation remains as strong as before, but the balance of forces in that linkage has changed substantially. Previously, politics provided a context for cooperation, limits on what could be done (for both political and technology-transfer-control reasons), and an occasional stimulus to pursue cooperative activities that might not otherwise have had sufficient budgetary priority (such as ASTP). Program managers constantly faced the reality that the political linkage could work to disrupt cooperative undertakings, as events in Afghanistan and Poland had during 1982-87.

Today, the U.S. desire to promote economic and political stability in Russia and to provide tangible incentives for positive Russian behavior in areas such as preventing proliferation of missile and other military technologies is a powerful engine behind cooperation. As a result, the United States has made unprecedented commitments of resources to Russia,[6] including large payments in exchange for Russian products and services, and it

---

[6] See "The Financial Dimension," later in this chapter.

is now willing to place Russian hardware and launch services on the critical path of keystone NASA projects, particularly the space station. Only a few years ago, the *Report of the Advisory Committee on the Future of the U.S. Space Program* opposed placing **any** foreign cooperative contribution in the critical path of U.S. projects,[7] and NASA managers had resisted allowing even such a long-time ally as Canada to play a similar role on the space station without extensive agreement provisions against default.

Recognizing the risks inherent in this situation, particularly given Russian[8] political and economic instability, NASA has sought to put arrangements in place to hedge against any Russian default on commitments. Generally speaking, in robotic space science and applications, Russian participation is not essential to specific projects, making contingency planning possible and cost-effective.

On June 17, 1992, a new civil space agreement was concluded at the first summit between President George Bush and Russian President Boris Yeltsin. Drafted and quickly agreed to in preparation for the summit, the agreement was substantially enabling and permissive rather than restrictive.[9] For the first time since 1977, it raised the prospect of cooperation in human spaceflight, including "Space Shuttle and Mir Space Station missions involving the participation of U.S. astronauts and Russian cosmonauts." For the first time, the agreement also foresaw cooperation in space technology and explicitly raised the possibility of "working together in other areas, such as the exploration of Mars."

The 1992 agreement sanctioned a very significant increase in activity across the entire range of cooperative space science and applications projects between NASA (the U.S. lead agency) and the Russian Space Agency (RSA), the Russian Academy of Sciences, and several other Russian agencies.[10]

In a joint statement accompanying the agreement, the two governments also agreed to "give consideration to" a specific exchange of astronaut-cosmonaut flight opportunities and to a Shuttle-Mir rendezvous and docking mission. Finally, the government announced that NASA would be giving a contract to a Russian enterprise, Scientific Production Organization (NPO) Energia, principally to study the potential use of the Soyuz-TM spacecraft as an interim crew-rescue vehicle for Space Station Freedom.[11]

On July 20, 1992, NASA Administrator Daniel Goldin and RSA General Director Yuri Koptev released a Memorandum of Discussion on talks held in Moscow, which elaborated on the understandings reached in June. The two agency heads also agreed to expand the JWG structure set up by the 1987 agreement by adding biomedical life-support systems to the JWG on Space Biology and Medicine and by creating a Mission to Planet Earth JWG to concentrate on earth science flight projects. They added study of a Russian-provided rendezvous and docking system to the NPO Energia contract signed in June and discussed the flight of U.S. instruments on a spare lander for the Russian Mars '94 mission.

---

[7] *Report of the Advisory Committee on the Future of the U.S. Space Program* (Washington, DC: U.S. Government Printing Office, December 1990), p. 8.

[8] For simplicity, "Russia" is used throughout this chapter to denote the United States' cooperative partner because the overwhelming majority of U.S. cooperative projects to date are with Russia. Where a general statement is made that does not apply as well to Ukraine, the distinction will be made clear. Where Kazakhstan is meant, it will be explicitly identified.

[9] Text of the 1992 agreement and subsequent implementing agreements are in appendix A.

[10] Summary tables describing cooperative activities approved by each of the six joint working groups and under way as of the end of 1994 are in appendix B.

[11] This role reversal from the 1984 U.S. proposal for a simulated space-rescue mission seems to have gone unremarked at the time.

Both the astronaut-cosmonaut exchange and the Mars '94 agreement were finalized on October 5, 1992, when Administrator Goldin and General Director Koptev signed agreements on human spaceflight and Mars '94 cooperation following meetings in Moscow.

Cooperation under the JWG structure has proceeded relatively smoothly since the signing of the 1992 agreement. The first Russian instrument to fly on a U.S. spacecraft, the KONUS gamma-ray-burst detector, was launched November 1, 1994, on the U.S. Wind spacecraft, part of the International Solar Terrestrial Physics (ISTP) Program. On December 16, 1994, NASA and RSA signed an agreement for the reflight of TOMS and for the flight of the third version of the Stratospheric Aerosols and Gas Experiment instrument (SAGE-III) on Russian polar-orbiting meteorological satellites. NASA views the Russian commitment to provide the launch, operations, and supporting science for SAGE, in particular, as a significant Russian contribution to the U.S. Earth Observing System (EOS) program. It was also agreed at the December meeting that Version 0 of the U.S. EOS Data and Information System (EOSDIS) will be interconnected with appropriate Russian counterparts.[12]

In 1993, the proposed cooperation on the Mars '94 mission was scaled down to the provision of a single U.S. instrument for each of the two landers, after the Russian developers of the spacecraft proved unable to accommodate a third lander on schedule; subsequently, reportedly because of budgetary, technical, and production difficulties, the Mars '94 launch slipped to 1996. At the June 1994 meeting of the U.S.-Russian Commission on Economic and Technological Cooperation (hereafter, for brevity, the Gore-Chernomyrdin Commission), the principals directed NASA and RSA to study "Mars Together," potential cooperative Mars-exploration options involving launches by each side during the 1998 and 2001 launch windows, and a concept for joint exploration of the Sun and Pluto, called "Fire and Ice." At the December 1994 Gore-Chernomyrdin meeting, the principals decided only to continue joint studies and "agreed that all such planning should take into consideration appropriate budgetary and financial constraints."[13]

The United States and Russia have continued to play key roles in the multilateral IACG, which is now occupied mainly with the ISTP Program, and Russia has joined the Committee on Earth Observation Satellites (CEOS), the most important multilateral coordinating body for Earth remote-sensing-satellite operators. Both countries are also key players in the International Mars Exploration Working Group (IMEWG).

For the most part, U.S.-Russian cooperation under the JWG structure has followed the established pattern of past NASA international cooperative projects—adherence to principles such as clean interfaces and general avoidance of technology transfers—but there has been one important departure. Even before the dissolution of the Soviet Union, U.S. officials recognized that some U.S. subsidy of Russian hard-currency expenses would be required to keep cooperation on track. More recently, NASA has found ways to provide limited injections of hard currency through writing small contracts for engineering-model hardware and services such as preparing interface-control documents. NASA program managers generally believe that cooperation is not currently possible without such stimuli, but they express a strong desire to return to the traditional, no-exchange-of-funds partnership model as soon as this is feasible.

---

[12] Private correspondence from Charles Kennel, NASA Associate Administrator for Mission to Planet Earth, to Ray Williamson of OTA, Feb. 16, 1995. In his letter, Kennel also noted that NASA will pay the marginal costs for integration and test for the SAGE flight and the TOMS reflight, expected to total $5 million to $6 million.

[13] U.S.-Russian Commission on Economic and Technological Cooperation, "Joint Statement on Aeronautics and Space Cooperation," Dec. 16, 1994, pp. 2-5. See also Peter B. deSelding, "Russian Woes Hampering Mars Project," *Space News*, pp. 2, 20, Dec. 19-25, 1994.

■ Human Spaceflight and the International Space Station

Background

In early 1993, President Bill Clinton ordered that Space Station Freedom be redesigned to reduce construction and operating costs.[14] In response, NASA formed a redesign team, including members named by its existing partners as well as NASA and industry participants, which developed a set of three options (A, B, and C) to fit within cost profiles provided by the White House. To be able to consider potential applications for Russian hardware in the revised design, NASA quietly brought in a small team of senior Russian engineers to serve as "resources" for the redesign process, but their inputs to the first phase, in the spring of 1993, were very limited.

In June 1993, President Clinton selected Option A (a scaled-down modular space station) with some elements of Option B (the design closest to Space Station Freedom), and he allowed three months for NASA's "transition team" to create a new, merged design. Again, the existing partners were involved directly in the redesign process, and an enlarged team of Russian "consultants" was much more actively involved than it was in the spring. On the diplomatic front, a series of contacts between NASA and RSA over the summer of 1993, and between the two governments, led to a White House announcement at the end of the first meeting of the Gore-Chernomyrdin Commission on September 2, 1993, that Russia and the United States foresaw Russia joining the space station partnership.[15] As an essential part of the package, the United States committed to pay $400 million over four years for Russian space hardware, services, and data in support of the joint spaceflight program leading to the development of the International Space Station.

On November 1, NASA and RSA agreed on an addendum to the September 7, 1993, Space Station Program Implementation Plan. The program set out in the addendum is organized into three phases. Phase One (1994-97) is fundamentally an expansion of the program agreed to in the Human Spaceflight Agreement of October 1992 into a program of seven to 10 shuttle flights to Mir[16] (see figure 3-3 and photo on page 51), as well as five medium- to long-duration flights on Mir by U.S. astronauts. Phase Two (1997-98) involves U.S., Russian, and Canadian elements and achieves the ability to support three people in 1998 with the delivery of the Soyuz-TM crew-rescue vehicle (see photo on page 52). Phase Three (1998-2002) completes assembly of the station, including European and Japanese components (see photo on page 53 and figure 3-4).[17]

In December 1993, a formal invitation to participate in the space station project was issued by the existing partnership and accepted by the Russians. Also in December, at the second meeting of the Gore-Chernomyrdin Commission, the Human Spaceflight Agreement was amended to provide for the full Phase One program, and an initial letter contract was signed to begin implementation of the $400 million commitment.

Since that time, a series of negotiations with the Russians and the existing space station partners has produced significant progress toward a new set of agreements governing the partnership. In June 1994, at the third session of the Gore-Chernomyrdin Commission, NASA Administrator

---

[14] For a detailed discussion of the evolution of the current design and Russian participation, see Marcia S. Smith, "Space Stations," Congressional Research Service Issue Briefs, Washington, DC, October 1994 (updated regularly).

[15] Formally, the two governments agreed only on the joint development of a program plan that would be the basis of a U.S. government decision and consultations with the other space station partners.

[16] Shuttle flight STS-60 in February 1994, involving the flight of cosmonaut Sergei Krikalev on a Space Shuttle mission, is formally also considered part of the Phase One program.

[17] The latest published manifest, dated Nov. 30, 1994, shows a total of 44 flights in the four-year construction period, of which 27 are to be Space Shuttle flights. Those totals do not include flights to rotate crews at the station or to resupply fuel and other consumables.

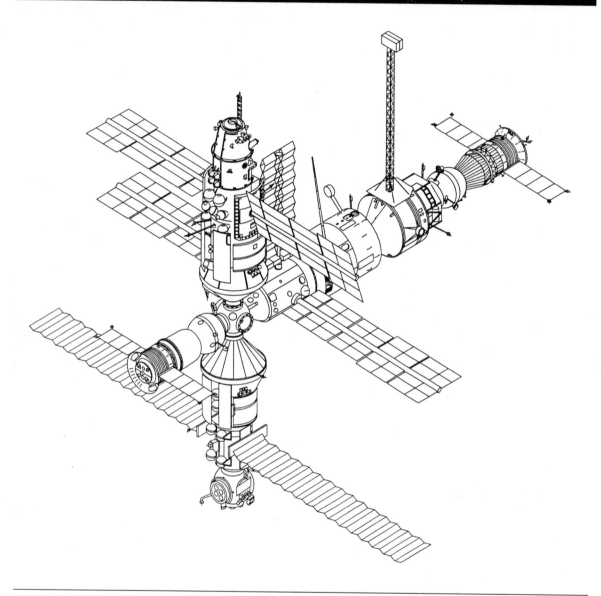

FIGURE 3-3: Mir Complex with Docked Progress-M and Soyuz-TM Spacecraft

SOURCE: David S.F. Portree, *Mir Hardware Heritage,* Houston, TX, 1994.

Goldin and RSA Director General Koptev signed an Interim Agreement covering initial Russian participation in the space station. The actual $400 million, fixed-price contract was also signed at that meeting. Negotiations are under way on a Memorandum of Understanding with Russia, on amending the existing Memoranda of Understanding with the other partners, and, in parallel, on amending the multilateral Intergovernmental Agreement to include the Russians and to bring it into conformity with the underlying bilateral agreements.

The original agreement structure stated that each partner would receive rights to use the space station proportionate to its contributions to the station, that each would pay the costs of its own assembly and logistics flights, and that the common operations costs would be shared among the

NATIONAL AERONAUTICS AND SPACE ADMINISTRATION

Artist's conception of U.S. Space Shuttle docked with Mir.

partners in proportion to each partner's contribution. The agreements envisioned that there would be a significant net flow of resources to the United States during the utilization and operations phase, which might be accomplished either through cash transfers or (preferably for the partners) through provision of goods or services. However, the very large Russian role in the station now includes elements formerly reserved for the United States (notably, provision of core systems and of transportation services during the assembly phase). ESA and Japan may become transportation providers as well. Negotiating allocations of space station resources and contributions to common operations costs is a challenging task; NASA hopes to complete the necessary negotiations and renegotiations during 1995.

Meanwhile, a series of milestones has been reached successfully in the development of the revised program. In particular, NASA and RSA reached technical and management agreements during August-September 1994, including a joint management protocol and an agreed specification document for the Russian segment of the space station. The first major shipments of equipment for use by U.S. astronauts on Mir were made in the September-December period, and the first top-level Joint Program Review was carried out in Moscow during November, confirming program milestones for 1995 and beyond. Rockwell In-

NATIONAL AERONAUTICS AND SPACE ADMINISTRATION

Artist's conception of Phase Two of the International Space Station.

ternational delivered the Space-Shuttle-to-Mir docking mechanism, incorporating key components from RSC Energia, to the Kennedy Space Center in Cape Canaveral, Florida, in November 1994. The Shuttle-Mir rendezvous and close-approach mission was successfully completed in February 1995, a key dress rehearsal for the docking missions to come. Finally, in mid-February, Lockheed, Khrunichev, NASA, and RSA successfully concluded separate, interlocking negotiations on purchase of the Functional Cargo Block (FGB) module, which provides guidance, navigation, and control capabilities for the Phase Two space station.

Progress has not been entirely smooth, however. Technical and organizational difficulties on the Russian side have been largely responsible for causing the scheduled date of the Spektr module's launch to Mir to slip from March until May 11, 1995. As a result, the first U.S. astronaut on Mir will have use of the equipment aboard for only about two weeks, rather than two months, as first anticipated; the next long-duration U.S. flight on Mir will not occur until March 1996.[18] In addition, severe problems with Russian customs clearance for the U.S. equipment involved in the flight have required the intervention of Vice President

---

[18] Part of the equipment is being launched to Mir on Progress cargo spacecraft instead.

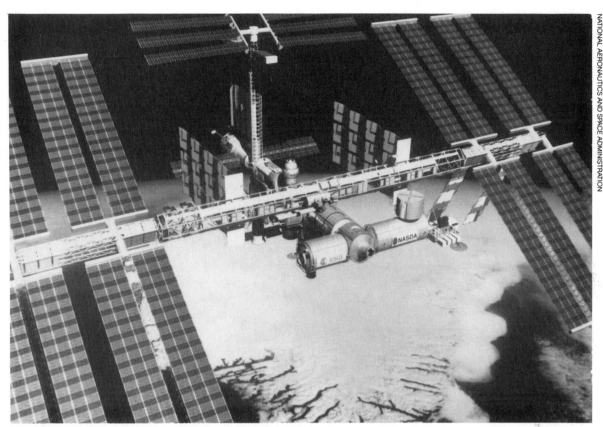

NATIONAL AERONAUTICS AND SPACE ADMINISTRATION

Artist's conception of Phase Three of the International Space Station.

Gore and Premier Chernomyrdin; a customs agreement was signed at the December 1994 Gore-Chernomyrdin meeting.

The Financial Dimension

NASA has historically conducted international cooperation on a no-exchange-of-funds basis. Since 1992, however, foreign policy and national security interests have led to a significant departure from this precedent in NASA activities with Russia. The effects of this change on NASA and on the place U.S.-Russian space cooperation occupies in the overall U.S.-Russian relationship are discussed in this section.

NASA payments to Russian entities, combined with directed procurements from Russian sources under NASA contracts with U.S. industry, will likely total nearly $650 million over the FY 1993-97 period:

- $400 million for space-station-related goods and services,[19]
- at least $210 million for the initial docking-mechanism purchase and the FGB procurements,
- $16 million for two Bion biosatellite flights, and
- at least $10 million in smaller procurements of goods and services.

---

[19] This comprises $335 million for Phase One Shuttle-Mir activities and $65 million for Phase Two, plus procurement of all Russian-provided docking mechanisms after the first one. As of March 1, 1995, only $62.5 million had been disbursed from the $100 million available in FY 1994; disbursements are made as deliveries of goods or services are received.

FIGURE 3-4: Schematic Drawing of the International Space Station

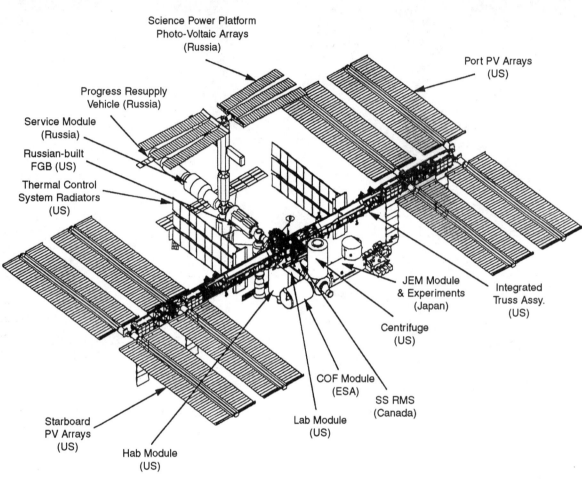

Science Power Platform
Photo-Voltaic Arrays
(Russia)

Port PV Arrays
(US)

Progress Resupply
Vehicle (Russia)

Service Module
(Russia)

Russian-built
FGB (US)

Thermal Control
System Radiators
(US)

JEM Module
& Experiments
(Japan)

Integrated
Truss Assy.
(US)

Centrifuge
(US)

COF Module
(ESA)

SS RMS
(Canada)

Starboard
PV Arrays
(US)

Lab Module
(US)

Hab Module
(US)

SOURCE: National Aeronautics and Space Administration, 1994.

These payments do not constitute **assistance** from NASA to RSA or to Russian space enterprises. The $400 million NASA-RSA contract covers at least seven Shuttle-Mir rendezvous and docking missions and up to 21 months of U.S. astronaut presence on Mir. NASA expects to gain fundamental experience in joint operations, including risk reduction, command and control, docking the shuttle with large structures in space, performing technology experiments, and executing a joint research program. The contract amount includes $20 million in support for jointly peer-reviewed

Russian scientists' proposals in all space-related disciplines and $25 million toward the cost of the FGB module being purchased by Lockheed from the Khrunichev Enterprise for use in the International Space Station. The FGB procurement by Lockheed, at a cost of $190 million, includes one unit and related services; NASA and RSA have agreed that RSA will contribute to NASA, at no cost, the FGB launch and all services not covered by the Lockheed contract, with the possible exception of some command-and-control software that may be needed.[20] The procurements of the docking mechanism, the Bion flights, and other, minor goods and services all involve the use of unique Russian capabilities by NASA at a low cost compared with the cost of developing them indigenously.

Nevertheless, no other executive branch agency is transferring funds to Russia at anything approaching this rate. U.S. government funds obligated for **assistance** to Russia through September 30, 1994, total something over $3 billion,[21] but over a third of that total is for in-kind goods (food shipments, principally in FY 1993), and significant funds that were obligated have been lost because of failure to spend them in time. Of the remainder, almost all have been paid to U.S. consultants and other entities to conduct assistance activities in Russia. Meanwhile, other non-NASA executive branch spending in Russia has been relatively minor.[22]

At the September 1993 Gore-Chernomyrdin Commission meeting, the United States committed $400 million of the NASA total payments to Russia when it agreed to involve Russia in the space station and to conclude an agreement on Russian access to the commercial space-launch market, in exchange for Russia's agreement to terminate its transfer of cryogenic-rocket-engine technology to India.

NASA funding is very important to the Russian space program. Inflation, the dramatic depreciation of the ruble, and conflicting data make it difficult to quantify this impact, but one senior RSA official said that RSA actually received R450 billion from the state treasury during 1994, about half its appropriation. Arguing for more state funding, he asserted that the total of all foreign agreements and contracts "represents just a fourth of our requirements."[23] However, at an average exchange rate of R3,000 = U.S.$1.00, the NASA/RSA contract alone yielded nearly R200 billion over that period.[24]

Aside from direct and indirect payments to Russian entities, NASA is committing significant budget resources to expenditures in the United States that are directly related to Russian cooperation. The totals stated by NASA in its FY 1996 budget submission are listed in table 3-1. Each item identified in the table is contained within broader program or project line items in the NASA budget, and some of the amounts in the table, such as the $100 million per year for "Russian Space Agency Contract," are included in the discussion of transfers to Russia above. In addition, the space station expenditures shown are

---

[20] Interview with Lynn F. H. Cline, Director, Human Space Flight Division, Office of External Relations, NASA Headquarters, Feb. 14, 1995.

[21] Office of the Coordinator for U.S. Assistance to the Newly Independent States, Department of State, "Cumulative Obligations of Major NIS Assistance Programs by Country to 9/30/94." See also U.S. Congress, Office of Technology Assessment, *Proliferation and the Former Soviet Union*, OTA-ISS-605 (Washington, DC: U.S. Government Printing Office, September 1994) for a discussion of nonproliferation-related U.S. spending programs involving Russia. This discussion includes Department of Defense funding under the Cooperative Threat Reduction Program.

[22] U.S. Congress, Office of Technology Assessment, op. cit., footnote 21, p. 28. Department of Energy (DOE) joint research programs with the Russian weapons laboratories are funded at $35 million in the FY 1994 Foreign Operations Appropriations Act, while the International Science and Technology Center (established to help fund Russian military scientists and engineers in civilian work related to their former fields) is funded at $25 million total, of which very little has been disbursed.

[23] Boris Ostroumov, Deputy General Director of the Russian Space Agency, quoted in "Manned Space Program in Imminent Jeopardy," Moscow *Trud*, in Russian, Dec. 10, 1994 (translated by Foreign Broadcast Information Service).

[24] If anything, this probably understates the impact because by the end of 1994, the exchange rate was approaching R4,000 = U.S.$1.00.

TABLE 3-1: NASA Russian-Related Activities Summary of Agency Programs and Costs with the Russian Republic ($ in millions—provided to Congress March 1995)					
	FY 1995	FY 1996	FY 1997	FY 1998	FY 1999
Russian Space Agency contract	100.0	100.0	100.0		
Mir missionsa	141.7	102.7	54.3	16.3	0.6
Space station-related developmentsb	20.0	20.0	10.0	0.0	0.0
Space science	14.4	10.1	9.2	12.3	6.2
Earth science	3.7	3.1	3.3	3.0	3.0
Space access	2.7				
Aeronautics	11.7	3.0			
Tracking and data	2.3	1.9	2.0	2.1	2.1
TOTAL	296.5	240.8	178.8	33.7	11.9

a Includes payloads and Shuttle/Spacelab support related to Mir and Shuttle-Mir missions.
b Does not include pending Lockheed contract costs.

SOURCE: NASA Headquarters.

subsumed within the $2.1 billion/year cap for space station spending.

PRIVATE SECTOR

U.S. private companies, for the most part, did not pursue potential business relationships in Russia or Ukraine until the demise of the Soviet Union. Since 1991, this situation has been changing, and cooperative efforts are beginning to bear fruit. In general, progress has been slow because of differences in business and technical cultures, as well as residual suspicions and restrictions left over from the Cold War.

■ Early Entrepreneurs and Glavkosmos: Before 1991

During the 1980s, a few small-scale entrepreneurial companies and individuals sought to open the U.S. market to Soviet launch services and remote-sensing imagery, with little success. Meanwhile, the Soviets formed Glavkosmos in 1985 as a mar-keting arm of their then-invisible Ministry of General Machine Building. Even earlier, there had been an abortive Soviet effort to commercialize the Proton launch vehicle, including requests that INTELSAT and INMARSAT, two international communications satellite operators, consider it as a candidate launch vehicle for their upcoming satellites. In this and subsequent efforts to qualify as a launch supplier for INTELSAT, however, Glavkosmos was unsuccessful.[25] Otherwise, little of consequence occurred during the late 1980s; one American firm successfully arranged for the flight of a small microgravity payload on the Mir Space Station in 1989, precipitating a brief but heated U.S. interagency dispute over whether the export of the experiment hardware had been properly approved.[26]

Several factors acted to limit the potential for private sector space business with the Soviet Union. First, Soviet secrecy about space-industry facilities and capabilities discouraged most companies from pursuing business ties; Glavkosmos

---

[25] After the collapse of the Soviet Union, Glavkosmos was reconstituted as a "private" company marketing space and other high-technology products and services. Although it continues to operate, the firm is not known to be involved in any of the major cooperative ventures currently under way.

[26] Another small payload was flown in 1992 without controversy.

was too obviously a front organization, and its officials were too abrasive and inexperienced in business. More important, all exports of space hardware and related technical data were controlled by the U.S. State Department under the International Traffic in Arms Regulations (ITAR). These regulations identified the Soviet Union and other Warsaw Pact countries as proscribed destinations, meaning that requests to export ITAR-controlled items to them were automatically denied unless a waiver of the proscription was approved at a high level in the Department of State, with the concurrence of other concerned agencies, most notably the Department of Defense (DOD).[27] Finally, most U.S. firms in a position to do business in Soviet space goods and services were heavily dependent on contracts with NASA and DOD; in the absence of clear, positive signals from these important clients, most firms chose not to pursue business ties in the Soviet Union.

■ Learning to Work Together: U.S. and Russian Industry

In the period following the breakup of the Soviet Union, the changed policy environment and the opening of Russian and Ukrainian enterprises to business contacts with the West resulted in a flood of Western aerospace business people to those countries. Initially, at least, some had hopes of acquiring space technology at "fire-sale" prices. Many went with authority only to visit, assess, and report back. The visitors found the Russian and Ukrainian aerospace sectors beginning slowly and painfully to abandon generations of secrecy and to learn Western business methods, while also confronting the devastating economic effects of dramatically reduced state contracts, hyperinflation, and a widespread breakdown of supplier and customer networks.

The Strategic Defense Initiative Organization (SDIO) actually initiated the first major "private sector" imports of Russian space technology beginning in late 1990, when it sought to import Topaz 2 space nuclear-reactor hardware and "Hall Effect" spacecraft thrusters (used for attitude control and station-keeping, or keeping the satellite in its proper orbit). SDIO used private firms as its purchasing agents for these procurements. Approval of these proposals by the U.S. government in March 1991, together with the decision to permit INMARSAT to negotiate with Russia for the Proton launch of a single INMARSAT satellite, signaled a significant shift in the U.S. government's attitude toward space trade with Russia.

Progress in developing business relationships has been slow, in most instances, and the Office of Technology Assessment (OTA) has not found any U.S. space enterprise that has yet shown a profit from its Russian activities. According to press reports and interviews conducted by OTA staff, the slow pace is attributable to factors on both sides. After the initial wave of U.S. "tire-kicking" visits, many Russian organizations felt that further contacts without tangible return were useless and began to reject further discussions unless the visitors could demonstrate, in advance, that they were prepared to invest substantial hard currency in the relationship. For their part, the Americans (and other Western businesspeople, as well) found the Russians often unwilling to provide financial and technical information that would have been a routine part of such exploratory exchanges in the West.[28] Even when business interest has been established and negotiations have begun, there have been serious conceptual and communications problems. Regulatory, legal, and bureaucratic obstacles on both sides disrupted schedules and strained relationships. Cultural differences, false

---

[27] Russia and the newly independent states continue to be proscribed destinations on the ITAR today.

[28] In part, this apparently reflected simple Russian inexperience; there also appear to have been significant residual security concerns and, in some instances, personal resistance to being asked to prove technical or managerial capabilities.

preconceptions, differing negotiating styles, and simple inexperience were all further complications. And always, there was the underlying political and economic uncertainty.

Nevertheless, U.S. firms are persevering and, in several areas, are increasingly optimistic about their prospects for the future.[29] The most promising prospects appear to be:

- *Marketing Russian and Ukrainian launch services, either from Russia or through innovative arrangements for launch elsewhere.* Lockheed is the firm most deeply involved, through the LKE International (Lockheed-Khrunichev-Energia) joint venture, but several others, including Boeing, are attempting to develop prospects involving Ukrainian launch vehicles and a variety of converted Russian missiles.[30]
- *Introducing Russian launch-vehicle and propulsion technology into U.S. systems through purchase and/or co-production arrangements.* Aerojet and Pratt and Whitney have each announced activities aimed at replacing the engines of existing U.S. launch vehicles; in addition it was announced at the Gore-Chernomyrdin Commission meeting in June 1994 that Pratt and Whitney would be working with NASA to explore the possible application of tri-propellant-rocket-engine technology developed by NPO Energomash, which might have application in future single-stage-to-orbit launch vehicles.
- *Marketing Russian remote-sensing-data products and services.* Firms including EOSAT, Worldmap International, and Core Technologies have announced the availability of Russian optical imagery with spatial resolution as good as 2 meters, as well as radar data from the Almaz satellites.

- *Using joint-venture efforts to apply Russian materials science and other underlying technologies to U.S. aerospace products.* Kaiser Aerospace and Electronics and McDonnell Douglas are among the firms pursuing these possibilities.
- *Using in situ Russian human resources in fields where their capabilities are well-known.* McDonnell Douglas, for example, has established joint research centers in Moscow and Huntington Beach, California, with the Mechanical Engineering Research Institute of the Russian Academy of Sciences, and it is pursuing a variety of technology and software development efforts.

■ Lessons

Cooperation to date in both the public and private sectors (including the experience of the Soviet period, although much more has been possible since 1991) has yielded a rich mix of lessons for the U.S. participants. OTA sought to collect and evaluate these lessons both through its November 9, 1994, workshop and through many interviews with people participating in cooperative activities. The following are the most broadly applicable principles that were identified by public and private sector managers:

1. *Although the payoffs can be great, and in some instances can only be gained through cooperation with Russia, cooperative activities with Russia are more difficult, take longer, and are, at this stage, riskier than is governmental cooperation with NASA's traditional partners or cooperation between U.S. companies and aerospace firms in Europe, Japan, and Canada. In some respects, the situation is comparable to the early stages of those established*

---

[29] A table listing representative private sector undertakings that have been reported in the press is in appendix C. Of course, important contacts are probably under way that have not been publicized.

[30] Daimler-Benz Aerospace of Germany and the Khrunichev Enterprise have recently announced a joint venture to market the Rockot space-launch vehicle, which is derived from the SS-19 intercontinental ballistic missile (ICBM) and can deliver small to medium-sized payloads to low Earth orbit (see Peter B. deSelding, "Rockot Launcher to Go Commercial," *Space News*, pp. 3, 6, Feb. 20-26, 1995).

relationships, but with a difference: the United States largely inculcated its space standards and practices in Western countries by virtue of its unchallenged leadership position during the 1960s and 1970s, but the Russian space infrastructure is already well-established and likely to resist changing its practices to conform to U.S. norms.

2. *There are wide linguistic, cultural, and societal differences between Russians and Americans, differences that are reinforced by the history of the past 75 years and the enforced separation of the U.S. and Russian space communities since the beginning of the Space Age. At the same time, technical people of the two sides tend to share an approach to the solution of problems in space technology and have a substantial body of common interest and mutual respect in space science.* Several consensus lessons follow from these basic observations:

- Whenever possible, understandings should be documented in detail, in writing, to avoid ambiguity. To remove as many misunderstandings as possible at the outset, it is very worthwhile to develop texts of important documents in both languages and to compare them formally and recognize both as equally authoritative.
- As one OTA workshop participant observed, Russia lacks settled legal frameworks for most business relationships, which are taken for granted in the West. As a result, it is important explicitly to define terms and establish agreement on the substance of contractual relationships, and not merely acceptance of language. Several workshop participants emphasized that Russian negotiators are quite willing to undo understandings reached earlier in order to exploit

political and time pressures to achieve their objectives.

- It is important to establish direct, open relationships and mutual respect based on technical competence. Russian society places great weight on personal relationships in business, particularly in the absence of established institutional structures for these new cooperative ventures. In addition, some U.S. participants believe that U.S. cooperation with Russia in space science has been more successful than ESA's or France's because, they say, Russian space officials recognize the United States as an approximate equal, while they regard other countries' space programs as inferior.
- Russian officials are extremely sensitive to any implication of condescension from the West, regardless of their currently weak economic position.[31]

3. *Even during the Soviet period, with plentiful resources and relative political stability, delays were frequently encountered in first-time scientific missions and original technological developments.* Conservatism in schedules is indicated; as one participant observed, schedules with no margin for slipping deadlines increase the risk of failure.

4. *Several workshop participants believe that internal bureaucratic conflict and disorganization are an important source of delay and disappointment to both sides.* They noted that proposed projects may well involve several Russian organizations, even if only the lead agency is represented in negotiations, and that these interagency relationships are in constant flux. Reliance on the principal Russian organization to deliver the others whose cooperation is needed can be risky because so little is

---

[31] One participant in OTA's workshop believes that the legacy of the 1980s has adversely affected current cooperative efforts by feeding a Russian perception that the United States is not serious about cooperation and seeks to take unfair advantage of Russia's current, disadvantageous position.

known about relationships among these organizations or their leaders.[32] Workshop participants and others also complained that officials and organizations on both sides continue to apply anachronistic controls on the transfer of space hardware and technical data, rather than acting to encourage the development of normal business relationships.

TOWARD NORMALIZATION

U.S. government program managers at the OTA workshop generally agreed that the large transfers of U.S. public funds to Russia currently being undertaken by NASA should not be continued longer than necessary (for either political or economic reasons). Several emphasized the desirability of developing a cooperative relationship with Russia that is comparable to those with the other major spacefaring nations.[33] Such a relationship would restore the principle of government-to-government cooperation with no exchange of funds (including an end to directed procurements across national boundaries). The cooperative element would be balanced by a vigorous commercial relationship involving an industry-determined mix of free and open commercial competition, on a reasonably level playing field, and teaming between U.S. and Russian firms where this makes business sense to the companies involved.

---

[32] Of course, such problems may be exacerbated when, for example, a project with Russia involves launches from Kazakhstan; the newly signed Russian-Kazakhstani agreement on the status of Baykonur may alleviate many of these concerns, but its implementation remains to be tested.

[33] One workshop participant believes that the United States should not seek to return to the general principles that govern its other cooperative relationships but should be willing to pursue a pragmatic, case-by-case approach (including fund transfers, as needed) for as long as necessary. This participant also believes that space science cooperation with Russia is dominated by unduly rigid adherence to such principles, and he praised the space station program's approach.

Other Countries' Space Cooperation with Russia | 4

T his chapter reviews the experience some other countries have had in space cooperation with the Soviet Union, and later Russia. It considers what lessons might be learned by the United States from their experience and addresses how the intensification of U.S. interactions with Russia in civil space efforts has affected and might in the future affect cooperative relations between the United States and its traditional partners in Europe, Canada, and Japan.

OTHER COUNTRIES' EXPERIENCE

Before the collapse of the Soviet Union in 1991, most of the other spacefaring nations had only very limited cooperative experience with the Soviet civil space program. The principal exception to this general rule is France, which opened space science cooperation with the Soviet Union under President Charles de Gaulle in 1966 and managed to maintain an active program in both robotic and, later, human spaceflight through the political vicissitudes of the 1970s and 1980s.[1] Since 1992, the European Space Agency (ESA) has joined France and the United States as Russia's main bilateral partners in civil space cooperation.

■ France and the Soviet Union

On June 30, 1966, French President de Gaulle and Soviet General Secretary Leonid Brezhnev signed the open-ended Intergovernmental Accord on Scientific/Technical and Economic Coopera-

---

[1] See appendix D for a French review of French-Soviet (later Russian) space cooperation. A thorough discussion of French-Soviet cooperation before 1985 can be found in U.S. Congress, Office of Technology Assessment, *U.S.-Soviet Cooperation in Space: A Technical Memorandum*, OTA-TM-STI-27 (Washington, DC: U.S. Government Printing Office, July 1985).

tion, which emphasized cooperation in the exploration and peaceful uses of outer space. Although it was intended as an assertion of French independence of action within the Western alliance, the agreement, and particularly its space component, soon acquired considerable substantive content. By the early 1980s, about one-third of the more than 2,000 space researchers and technicians in France were working in some way with French-Soviet cooperation in space, and the level of French funding for cooperation with the U.S.S.R. was not far below that for cooperation with the United States. This balance was an apparent, and relatively explicit, objective of the French program.[2]

Since those early days, French bilateral space cooperation with the U.S.S.R. has remained concentrated in a few areas, notably:

- astronomy and astrophysics,
- space plasma physics,
- planetary exploration,
- materials processing in space, and
- life sciences.

The French have invested significant resources in cooperation with the Soviet Union in planetary exploration. The Vega mission, launched in 1984 to explore Venus and Halley's Comet, featured French-built atmospheric balloons that were successfully released and tracked in the Venusian atmosphere in 1985. Similar—but more sophisticated—French balloons are intended as part of the next Russian Mars mission, recently postponed to 1996. Major French instruments also flew on the Soviet Granat and Gamma missions in 1989 and 1990.

In addition, in 1982, France and the Soviet Union began a series of cooperative human-space-flight activities with the flight of Jean-Loup Chretien on Salyut-7. After the flight of Patrick Baudry, Chretien's backup, on the U.S. Space Shuttle in June 1985 (both the United States and France were apparently seeking balance in this high-profile field), Chretien flew again in 1988 aboard Mir and conducted the first French EVA (extra-vehicular activity, or "spacewalk"). Another French "spationaut" flew on Mir in 1992.

In December 1989, the French and Soviets signed a long-term agreement on human-space-flight cooperation, calling for a series of flights on a reimbursable basis, in 1993, 1996, 1998, and 2000. Most recently, plans to shut down Mir in late 1997 or early 1998 appear to put the later flights in jeopardy, but negotiations continue, with the price for the 1996 flight quoted as $13.7 million.[3]

From a U.S. policy perspective, the most interesting aspect of the conduct of French-Soviet space cooperation is the difference between the U.S. and French responses to past changes in the political environment. While the United States allowed its intergovernmental space agreement with the Soviet Union to lapse in the wake of the Soviet invasion of Afghanistan and the imposition of martial law in Poland, the French decided to continue relations. Indeed, Chretien's Salyut-7 flight in 1982 was the subject of considerable controversy in France, but the issue appears to have been resolved in favor of continuing cooperation at a higher or lower profile, depending on the political environment, rather than suspending ties.[4]

Since the French lacked a crewed spacecraft of their own, as well as the resources for an extensive flight program in space science, a decision to terminate cooperation with the Soviet Union would have been comparatively costly. Moreover, as noted above, independence and balance between the United States and the U.S.S.R. were important tenets of French foreign policy in the 1980s. The French also saw space cooperation as important in working toward broader objectives such as im-

---

[2] Ibid., p. 54.

[3] Peter B. DeSelding, "French Try for Mir Swan Song," *Space News*, p. 21, Jan. 9-15, 1995.

[4] U.S. Congress, Office of Technology Assessment, op. cit., pp. 61-66.

proved communications and reduced tensions between the U.S.S.R. and the outside world. Finally, they appear to have judged that their systems for controlling and monitoring technology transfers in the course of cooperative projects were sufficiently effective to obviate any concern about unwarranted transfers of militarily significant items.[5]

■ The European Space Agency

Although there was some scientist-to-scientist contact between European and Russian space scientists during the 1980s, for all practical purposes, ESA's engagement with the Russian Space Agency (RSA) began in 1991, when, at an ESA ministerial meeting in Munich, ESA decided to explore the potential for expanded cooperation in support of its human-spaceflight objectives. This decision was confirmed during the subsequent ministerial meeting in Granada, Spain, in November 1992.

During 1993, ESA and RSA established working groups to focus on five areas of human space-flight:

1. European astronaut missions on Mir,
2. in-orbit infrastructure,
3. crew and freight transportation vehicles,
4. space-transport-systems technology, and
5. in-orbit servicing.

In addition, in mid-1993, ESA and the Russian enterprise NPO Energia signed a contract, later confirmed by an ESA-RSA agreement in October 1994, covering paid flight of European astronauts on Mir. The first flight, a 31-day mission, took place in October-November 1994, and a 135-day flight (Euromir '95) is scheduled to begin in August. The latter flight will include a spacewalk.

Early technical exchanges concerning reusable spacecraft for human spaceflight and space suit design have not been pursued, but with the deci-

sion to involve the Russians in the International Space Station, ESA has begun negotiations with RSA in two areas directly related to that project: providing a data-management system for the Russian service module and providing the European Robotic Arm (ERA) for installation on the exterior of the module (see figure 3-8). Terms of the Memoranda of Understanding governing these activities have not yet been finalized, but in return for providing the ERA, ESA will benefit from its qualification for and use in space, while the quid pro quo for the data-management system will probably be in the form of Russian space hardware, reportedly including the docking mechanism Russia currently uses to attach its station modules to the Mir core.

ESA is dedicating significant resources to this cooperative initiative. Its budget for the European astronaut flights on Mir is $82 million. Within Europe, ESA is spending approximately $60 million on the data-management system, and it decided in September 1994 to spend $180 million for the ERA.[6]

From November 1992 through the end of 1994, ESA committed to pay a total of about $81 million to Russian entities.[7] Of this amount, $56.4 million funds the contract with NPO (now Russian Space Corporation (RSC)) Energia, which is responsible for Russian implementation of the astronaut flights on Mir and payment of any subcontractors. Another $6 million was approved to reimburse RSA for the flight of ESA payloads on Russian Foton recoverable spacecraft.

European budgetary difficulties are putting strong constraints on ESA's ability to expand work with Russia, however. During 1994, ESA was considering proposals for cooperative development with Russia of a crew-return vehicle (CRV) for the space station (which could evolve into a crew-transfer vehicle to carry crews to and from orbit) and an automated transfer vehicle

---

[5] Ibid., p. 66.

[6] Peter B. deSelding, "Ventures with Russia Starting To Bear Fruit," *Space News*, pp. 10, 17, Oct. 31-Nov. 6, 1994.

[7] Figure provided by Karin Barbance, Russian Desk Officer in the International Affairs Department, ESA Headquarters, Mar. 2, 1995.

(ATV) upper stage to deliver Ariane 5 payloads to the station. ESA was reportedly also considering options for joint development of the CRV with the National Aeronautics and Space Administration (NASA). Recent news reports suggest that ESA may scale back its contribution significantly and is not actively pursuing the Russian option.[8]

EFFECTS OF U.S.-RUSSIAN PACTS ON OTHER U.S. PARTNERSHIPS

The dramatic expansion in U.S.-Russian space cooperation since 1992 has taken place in the broader context of space relations between the United States and traditional partners. Those relationships have concurrently been undergoing fairly significant change in their own right, as the United States and the partners all reassess their space plans in the face of tight budgets and shifting national priorities.[9] This section briefly examines the impact of U.S.-Russian developments in various areas on the United States' cooperative relations with other nations and international organizations.

■ Space Station

Since the signing of the Intergovernmental Agreement on Cooperation in the Detailed Design, Development, Operation, and Utilisation of the Permanently Manned Civil Space Station and its companion Memoranda of Understanding in September 1988, the partners have preserved their cooperation and made significant progress, though the course has not been smooth.[10] As negotiated, those agreements provide for NASA's clear preeminence in the program, commensurate with its provision of the core space station and supporting infrastructure for all of the partners' contributions. For their part, the partners had sought equality in the program's decisionmaking process but settled for a commitment by all parties to seek consensus; final authority, in the absence of consensus, was reserved to NASA.

Through a series of design reviews and redesigns driven by U.S. budgetary and political forces, NASA tried, with varying success, to balance its domestic needs with consideration for those of the partners. In 1989, an internal NASA design review was initially concealed from the partners, leading to a stormy consultative meeting at the governmental level in September of that year. In subsequent restructuring and redesign exercises, NASA made considerably more effort to involve and consult with the partners. For their part, the partners generally accommodated the resulting design changes, but at a price, in terms of schedule changes and increased costs.

The Clinton Administration's 1993 decisions to redesign the space station dramatically and involve Russia in a key role sharply increased tensions in the partnership. From the partners' perspective, throughout the 1993 redesign and transition process, the United States failed adequately to consult its partners. When President Clinton went "over the heads" of the space agencies and wrote to his counterparts in Europe, Canada, and Japan in October 1993, seeking their support for inviting Russia to join the project, he further exacerbated the resentment of partner space agencies. However, if he had not interceded, it is by no means clear that the space agencies would have reached agreement on Russian partici-

---

[8] See Craig Covault, "Europe Faces Critical Decisions on Station Role," *Aviation Week & Space Technology*, pp. 22-23, Jan. 16, 1995; and Peter B. deSelding, "European Outlook Cloudy," *Space News*, pp. 8, 10, Feb. 13-19, 1995.

[9] A forthcoming OTA background paper, *International Collaboration in Large Science and Technology Projects*, examines trends in this and other key areas of large-scale international science and technology cooperation.

[10] See Marcia S. Smith, "Space Stations," Congressional Research Service Issue Brief 93017, Washington, DC, October 1994 (updated periodically). Also see Graham J. Gibbs, *Expanding the International Space Station Program Partnership—An International Partner's Perspective* (presented at the 45th Congress of the International Astronautical Federation, Jerusalem, Israel, Oct. 9-14, 1994).

pation, at least in time for the December 1993 announcement that the United States desired.[11]

NASA now believes that equilibrium has been restored in the relationship. Cooperative activities are proceeding well, relations with partner representatives are cordial in Washington and Houston, and the negotiations for revised Memoranda of Understanding and an amended Intergovernmental Agreement, although substantively difficult, are proceeding relatively smoothly.[12]

After a series of difficult, ministerial-level decision meetings, ESA announced in 1993 that it was reviewing the scope and character of its contribution to the International Space Station and set a decision point in early 1995. Although the ESA Executive has produced a series of detailed and varyingly ambitious plans for a redefined space station commitment, it has not yet decided how to proceed. France has declared that it will be unable to reach a decision at the ministerial level until October 1995, seven months after NASA says it must have ESA's decision.[13] At French and German insistence, the ESA Executive (the administrative staff in Paris) circulated an "alternative scenario" to member states early in February that seeks to reduce spending between 1996 and 2000.[14]

In early 1994, Canada informed the United States that it would have to withdraw from the space station program unless a means could be found to reduce the cost of its contribution. Detailed and painstaking negotiations resulted in an acceptable restructuring, reducing Canadian costs by approximately U.S.$550 million and securing a Canadian recommitment.[15]

Some of the difficulties in Europe and Canada result from a general decline in support for space spending, particularly spending on human spaceflight. There is no doubt, however, that partner resentment over the U.S. management of Russia's entry into the program did political harm. More broadly, the space station experience appears to have convinced the partners that they should not enter into such an asymmetrical arrangement again.[16] It is not yet clear whether, or to what extent, this determination will hamper efforts to renegotiate the space station agreements by the end of 1995, as NASA now plans.

▌ Space Science and Applications

The situation for collaboration in space science and applications is considerably different from that for space station collaboration. Reasons for this difference include:

- There has been a strong tendency toward increasing multilateralism in space science since the founding of the Inter-Agency Consultative Group for Halley's Comet (IACG) in the early 1980s. Russian scientists and managers were involved from the group's inception. In remote sensing, a variety of multilateral mechanisms has existed since the 1960s and 1970s to coordinate remote-sensing-program plans and policies. In 1993, NASA and the National Oceanic and Atmospheric Administration (NOAA) were successful in securing Administration approval for a U.S. initiative to invite Russia to become a member of the Committee on Earth

---

[11] Gibbs, op. cit., pp. 3-6; Gibbs notes, in particular, that although NASA involved its existing partners in the 1993 redesign and transition processes, leading to adoption of a redesigned space station, that process did not explicitly anticipate Russian participation. Instead, the United States and NASA negotiated with the Russians on a bilateral basis, only informing the partners on the eve of the September 1993 meeting of the Gore-Chernomyrdin Commission meeting.

[12] Interview with Lynn F. H. Cline, Director, Space Flight Division, Office of External Relations, NASA Headquarters, Feb. 14, 1994.

[13] Peter B. deSelding, "Europe, U.S. Scramble for Station Funds," *Space News*, pp. 3, 20, Jan. 16-22, 1995.

[14] Peter B. deSelding, "ESA Makes Cuts, Delays to Space Station Pledge," *Space News*, p. 3, Feb. 6-12, 1995.

[15] Canadian Space Agency press release, June 3, 1994.

[16] Gibbs, op. cit., footnote 10, p. 3. In particular, the partners believe that decisionmaking mechanisms that give the United States the last word are inconsistent with true partnerships.

Observation Satellites (CEOS), a key international body in the field. Current U.S. policy, in both space science and earth science, is to encourage further Russian integration into worldwide, coordinated activities.

- Russian emergence as a significant player has not undone existing arrangements, as happened in the space station program. Indeed, the Russians pioneered, in some respects, the beginnings of multilateral space cooperation through the science working groups established in the 1980s for their planetary and astrophysics missions, which relied heavily on foreign instruments.

- Programs in these areas generally have a lower political profile than those in human spaceflight. Although they are no less vulnerable to the annual budget process, they are less obvious captives to linkage with the overall political climate.

■ Commercial Relations

In general, business relationships among U.S. and Russian firms have developed without unduly affecting either side's relations with third parties in Europe, Japan, and elsewhere. The one potential exception to this rule is trade in launch services. In 1989 and 1993, respectively, the United States agreed to the entry of China and Russia into the world market for commercial launch services.[17] To guard against market disruption caused by the entry of nonmarket launch-service providers, the United States negotiated launch trade agreements with each country that provided quantitative limits on the number of launches each could provide and specified pricing controls intended to prevent artificially low bids.

Recently, the United States has renegotiated its launch trade agreement with the People's Republic of China, giving the Chinese a significantly larger quota and more leeway on price than that afforded the Russians in the 1993 commercial space-launch agreement with them.[18] There have been hints that the United States may consider lifting the quantitative restriction on commercial sales of Russian launch services altogether.[19] Such an action, in response to Russian and U.S. urging, could have a major impact in Europe. The European firm Arianespace is already critical of what it sees as the United States' failure to enforce the price requirements of the 1993 agreement.[20] Liberalization or elimination of the U.S.-Russian agreement might be seen in Europe as a blatantly anti-Arianespace move by the United States, particularly if NASA and Department of Defense launches continued to be reserved for U.S. launchers only.

---

[17] Previously, the United States had been able to block such entry by denying export licenses for satellites or satellite components; all commercial satellites built outside the United States included U.S. components, so this restriction was effective.

[18] Warren Ferster, "China Wins Big In Launch Deal," *Space News*, pp. 1, 20, Feb. 6-12, 1995.

[19] "Russia and US May Scrap Commercial Russian Rocket Launch Quota," *Interfax*, Moscow, Jan. 27, 1995 (translated by the Foreign Broadcast Information Service).

[20] Warren Ferster, "Russia: Relax Launch Limits," *Space News*, Dec. 19-25, 1994, p. 1; Andrew Lawler, "Industry Criticizes U.S. Launch Agreements," *Space News*, p. 3, Oct. 3-9, 1994.

Opportunities for and Impediments to Expanded Cooperation | 5

F oreign policy considerations, together with the budgetary pressures currently facing the civil space programs of the United States and of other spacefaring nations, provide a strong motivation for examining whether expanded space cooperation with Russia is desirable, in what fields, and on what basis. Additionally, there is a need to examine more closely the risks that would go along with such expanded cooperation and how those risks might be mitigated; this discussion is relevant to current cooperative programs as well. Finally, this chapter examines the role of the Russian and U.S. governments in civil space cooperation, particularly regarding control and regulation of private sector activities.

POTENTIAL FOR EXPANDED COOPERATION[1]

■ Launch Vehicles and Propulsion

Although the Clinton Administration's National Space Transportation Policy[2] directs the U.S. government to negotiate and implement agreements controlling trade in commercial space-launch services, it expressly authorizes the use of foreign launch services on a no-exchange-of-funds basis for cooperative government-to-government programs. The policy also states that "the U.S. Government will seek to take advantage of foreign components or technologies in upgrading U.S. space transportation sys-

---

[1] As used in this chapter, the term *cooperation* encompasses both government-to-government relationships and private sector ties such as joint ventures, co-production, and long-term supplier relationships.

[2] The White House, Office of Science and Technology Policy, "Fact Sheet—National Space Transportation Policy," Aug. 5, 1994.

| 67

tems or developing next generation space transportation systems."

The policy clearly was framed, among other things, in the knowledge that the greatest strength of the Russian space program (and the principal strength of the Ukrainian program) lies in launch vehicles and associated technologies, particularly propulsion and rapid payload processing and integration. The availability, robustness, and established reliability of Russian and Ukrainian launch vehicles, built on large-volume series production over many years, are potentially major assets for cooperative civil space activities. The use of those launch vehicles on a no-exchange-of-funds basis could permit some missions that would not be undertaken otherwise. Private sector development of these capabilities could also be a significant economic asset for Russia and Ukraine, but this dimension is currently limited by Western unwillingness to allow those states full access to the launch market.

As a practical matter, use of Russian and Ukrainian launch vehicles is being pursued on several fronts:

- Russian launch vehicles are being extensively scheduled to provide critical transportation for the assembly and operation of the International Space Station.
- The U.S. and Russian governments are discussing the use of Russian launch vehicles in cooperative projects such as planned missions to Mars and Pluto.
- The Lockheed-Khrunichev-Energia joint venture (LKE International) is marketing Proton launch services internationally, for both geostationary and low-Earth-orbit satellites.
- Boeing is seeking U.S. government approval for a joint venture with Ukraine's NPO Yuzhnoye (also known by its Ukrainian name, NPO Pivdenne),[3] RSC Energia, and a Norwegian builder of offshore oil platforms to market launch services using the Zenit vehicle.[4]
- U.S. manufacturers of propulsion and launch vehicles are pursuing proposals for the use of Russian propulsion systems, both to re-engine existing U.S. launchers and to include in proposals for future systems such as the X-33 reusable-launcher demonstration vehicle.

Space station program planners anticipate that Ukrainian Zenit launch vehicles (with Russian main engines) will be a key transportation element in the space station project, and a Russian-Ukrainian agreement is being negotiated to cover the provision of these and other Ukrainian goods and services to Russia for its use in the project. The agreement was expected to be ready for signature in 1995, but recent press reports suggest that the negotiations are not going well and that Russia is seeking to reduce its dependence on Ukrainian suppliers.[5] Meanwhile, the Boeing joint-venture proposal awaits licensing by the U.S. Department of State. Press reports indicate that the U.S. government is withholding its approval in part out of concern for the impact of another new entry in the commercial space-launch market, and also as leverage to help ensure Ukrainian conformity with the Missile Technology Control Regime. The regime seeks to deny the transfer of systems capable

---

[3] An umbrella space agreement between the United States and Ukraine was signed November 22, 1994, by Presidents Bill Clinton and Leonid Kuchma. The agreement is very similar to the 1992 U.S.-Russian agreement and is permissive rather than specific. The two Presidents also announced that NASA and the Ukrainian National Space Agency will prepare recommendations for flight of a Ukrainian payload specialist on the Space Shuttle (see "Joint Statement on Future Aerospace Cooperation Between the United States and Ukraine," Office of The Press Secretary, The White House, Nov. 22, 1994).

[4] The Zenit uses a highly automated launch-processing system, which could give it a competitive advantage; this Russian innovation could, in principle, be applied to evolving U.S. systems, as well.

[5] Peter B. deSelding, "Russia Ready To Use Ukraine-Built Zenits," *Space News*, pp. 1, 21, Oct. 3-9, 1994; "Zenit Rockets To Be Used in International Space Project," Kiev *Unian* (in Ukrainian), Nov. 15, 1994 (translated by the Foreign Broadcast Information Service); Peter B. deSelding, "Russia Distances Space Program from Ukraine," *Space News*, p. 3, Feb. 20-26, 1995.

of the long-range delivery of weapons of mass destruction.[6]

■ Spacecraft

Russian spacecraft capabilities are mixed. The fact that the Russians use robust, simple, low-cost, shorter-lived operational systems does not indicate, as some have argued, that they are necessarily inferior to U.S. designs—simply different. This difference does mean, however, that it may not be cost-effective to adapt high-cost, unique U.S. instrument designs, developed for long-lived U.S. spacecraft, to fly on Russian operational spacecraft with a shorter lifetime. Simpler instruments, or instruments that replicate existing hardware, may be a good fit, however, depending on the cost of adapting them to the new platform.

In the past, the United States has not been able to anticipate some adaptation costs. For example, in preparing to fly the Total Ozone Mapping Spectrometer (TOMS) instrument on a Russian Meteor-3, the National Aeronautics and Space Administration (NASA) initially assumed that the Russian satellite included a mass data storage subsystem, as is standard in U.S. satellites. Instead, NASA learned that the Russian satellite did not have this capability, which meant that NASA had to build and fly, for the first time, a solid-state memory unit dedicated to TOMS data. In addition, NASA learned that Russian meteorological satellite integration facilities did not have "clean room" capabilities for protecting sensitive instruments from contamination; because the TOMS required such handling, NASA provided a portable "clean room" for the TOMS integration.[7]

Russian scientific spacecraft present somewhat different issues. Some, such as the Luna and Venera planetary probes, were robust and resilient designs. Others, notably the two ill-fated Phobos spacecraft, both of which failed after launch in 1988, were not.[8] Some U.S. spacecraft specialists attempting to understand Russian spacecraft designs have had difficulty with the withholding of specific design information by the spacecraft manufacturers—at least in part a residuum of Soviet-era secrecy and bureaucratic compartmentalization, as well as a reflection of the Russians' perception that the design information might have commercial value.

Maintenance of schedule on new-design scientific spacecraft has also been a Soviet (and now a Russian) weakness; historically, the Russians have been much more successful at producing a series of spacecraft once a design is in series production. The current delay in completing the Mars '94 spacecraft, for example, reportedly results as much from technical problems as financial shortfalls.[9]

Mir and Mir-related spacecraft (such as the Functional Cargo Block (FGB) and other major Russian components of the space station) represent a special case. On the one hand, Russian experience in human spaceflight is unmatched, and Mir systems, although not technologically as sophisticated as systems being planned in the West for use on the space station, are mature and well tested. On the other hand, the FGB has not flown in the form that will be required for the space station, and delivery-schedule problems have been common during Mir's lifetime—the Spektr and

---

[6] See, e.g., Warren Ferster, "U.S. Eyes Zenit Warily," *Space News*, pp. 1, 28, Dec. 12-18, 1994.

[7] From an unpublished interview with George Esenwein, NASA Program Manager for the TOMS/Meteor-3 flight, 1991.

[8] One actually failed due to an erroneous command from the ground, but it was observed at the time that the spacecraft lacked fail-safe provisions that might have enabled controllers to save the mission.

[9] Frank Morring, Jr., "NASA Applies New Philosophy To Meet Old Goals in Mars Exploration," *Aerospace Daily*, p. 111, Oct. 21, 1994.

Priroda modules, for example, were originally scheduled for delivery in the late 1980s, then in 1992, but will not reach Mir until this year.[10]

As discussed in chapter 2, there are basic differences in spacecraft and instrument design philosophy between the U.S. and Russian programs, which can make designing and implementing interfaces between U.S. and Russian hardware difficult. U.S. systems have tended to be expensive, complex, high-performance, long-lived, heavily ground-tested, one- or few-of-a-kind designs. The Russian approach, on the other hand, emphasizes relatively low-cost, simple, moderate-performance systems that are flight-tested and put into series production, with the expectation that they will need to be replaced on orbit in a comparatively short time.[11]

■ Instrumentation

Russian scientific spacecraft, particularly during the 1980s and early 1990s, relied heavily on foreign instrumentation.[12] Western instrument technology is generally acknowledged to be superior. On the other hand, the Soviet Union had outstanding success in such technological areas as automated sample return (the Luna series), automated roving vehicles (the Lunokhod), and the series of Venera landers in the high-temperature, high-pressure environment of the Venusian surface. Other instruments and components with military applications or ancestry (such as the Vega imaging system, which used a Soviet military

charge-coupled-device (CCD) array in a Hungarian-designed camera with French optics) have been very successful, as well. Russian military-derived remote-sensing systems, particularly those using photographic film, also produce excellent results. Radar-imaging systems with a similar heritage may be another asset, and there are reportedly plans to commercialize high-resolution, digital optical-imaging systems in the near future.

■ Human Resources

According to NASA officials and other observers, Russian scientific and engineering talent represents a great strength. Russian capabilities in mechanical engineering, software development for science and engineering, and science theory are excellent.[13]

"Brain drain" represents a major potential problem for Russia, as the best (or best-known) specialists are offered opportunities to leave for jobs outside Russia or in other fields. One reason for the U.S. government to support programs that stress in situ employment of such people is to counter such losses of talent, both to stem potential proliferation of militarily relevant know-how abroad and to encourage economic development and defense conversion at home.[14]

■ Other Capabilities

Russian deep-space communications assets—notably, the 70-meter-class antennas at Yevpatoria

---

[10] A launch schedule for 1995 Shuttle-Mir activities, including launch of the two modules, was signed by Russian Space Agency (RSA) General Director Yuri Koptev and NASA Administrator Daniel Goldin at the December 1994 meeting of the Gore-Chernomyrdin Commission. This came after a late-1994 announcement of a two-month slip in the Spektr launch date, which seemed to threaten the scientific return on the investment in NASA astronaut Norman Thagard's visit to Mir, beginning in March 1995.

[11] Some observers have suggested that these differing design philosophies also reflect fundamental systemic differences in approach to technology. In this view, capitalist firms tend to look for new technological solutions and invest heavily in research and development, while Soviet (now Russian) entities place a lower value on innovation and seek to make the best (often clever) use of existing technology rather than take the risk of developing something new.

[12] Much of the major instrumentation on the Vega and Phobos missions, for example, was of Western European origin.

[13] An early initiative under the 1987 space agreement was the exchange of scientists between the science teams of various missions, including Phobos, Magellan, Mars Observer, and Cassini.

[14] See U.S. Congress, Office of Technology Assessment, *Proliferation and the Former Soviet Union,* OTA-ISS-605 (Washington, DC: U.S. Government Printing Office, September 1994), pp. 62-66, and chapter 6 of this report.

(actually in Crimea, Ukraine, but controlled by Russia) and Ussuriysk—could provide a useful complement to the capabilities of NASA's Deep Space Network, particularly in times of extremely high workload such as during the Galileo encounter with Jupiter beginning in December 1995. In 1992, the first NASA contract awarded to a Russian entity involved feasibility studies of such complementary uses, which demonstrated that using the Russian antennas would be of some modest value. Negotiations broke down, however, when NASA's Russian counterparts demanded a price for the use of the antennas that was much higher than NASA was prepared to pay. Discussions were broken off at that point, in 1993, and NASA has since developed and implemented other plans to handle the expected workload.[15]

Russian capabilities in advanced materials offer a potential for commercial development that has not, so far, been realized. U.S. engineers have explored the use of materials such as aluminum-lithium alloy, titanium, and carbon-carbon composites on U.S. spacecraft and launch vehicles. Kazakhstan has significant production capacity for beryllium, but a joint-venture project to exploit this capability, which was launched soon after the dissolution of the Soviet Union, has achieved only limited results.[16]

RISKS AND RISK MANAGEMENT

Clearly, civil space cooperation with Russia involves risks—some that are common to cooperation with the United States' traditional cooperative partners and some that are unique to Russia. This section characterizes those risks and discusses some options that managers in both the United States and Russia might adopt for managing them.

∎ Risks

The programmatic benefits of international space cooperation are offset, to some degree, by:

- an increment of technical risk (presuming that the international partner's technical capabilities are not as well known as one's own, or that new developments are required);
- added management complexity; and
- exposure to additional political risk, if only because the needed budgets must survive two or more political systems rather than only one.

Generally, NASA and its traditional partners have judged these risks worth taking.[17]

In the case of cooperation with Russia, the picture is somewhat more complex, and additional risk factors are clearly present. These additional factors include:

- Russian political and economic uncertainties on the most fundamental level, which cast doubt on whether (or when, at least) commitments will be honored, whatever the intentions of the parties.
- The risk of "reverse linkage," in which strains in other aspects of the U.S.-Russian relationship adversely affect cooperative space projects (this has happened before, most clearly in 1982, when the government-to-government space agreement was allowed to lapse to express U.S. ire over the imposition of martial law in Poland).
- Russian systemic immaturity, that is, the substantial lack of a settled legal and institutional framework within which cooperation can go forward in a relatively predictable fashion.
- Exacerbated programmatic uncertainties, deriving from limited cooperative experience and 30 years' mutual isolation.

---

[15] Interview with Charles Force, NASA Associate Administrator for Space Communications, Dec. 22, 1994.

[16] See appendix C.

[17] The overall record in high-technology cooperation with U.S. friends and allies is distinctly mixed, principally because of institutional mismatches (e.g., annual funding in the United States versus multiyear funding in other countries). See U.S. Congress, Office of Technology Assessment, *Arming Our Allies: Cooperation and Competition in Defense Technology*, OTA-ISC-449 (Washington, DC: U.S. Government Printing Office, May 1990).

- Reliance on the Baikonur launch site in Kazakhstan, with its attendant political and infrastructural uncertainties.
- Problems of communication and understanding, again derived from a lack of common experience and from cultural factors.

■ Risk Management

Political and economic uncertainties in Russia (and elsewhere in the former Soviet Union) present particular difficulties for risk management in civil space cooperation, as described in chapter 2. During the Office of Technology Assessment's (OTA's) November 9, 1994, workshop, "Civil Space Cooperation with the Former Soviet Union," several participants expressed doubt that the deteriorating overall condition of the Russian aerospace sector will permit it to deliver on the commitments to space cooperation being made by the Russian government. The Ukrainian economy, including its small aerospace sector, is in even worse condition than Russia's. Under these circumstances, it is extraordinarily difficult for U.S. program managers to decide how much to invest in hedging against the Russian (or Ukrainian) partner's default. The extent of such hedging is likely to be limited by available resources on the U.S. side, but some increment of confidence could be gained through further systematic analysis of post-1991 trends in the Russian aerospace sector.

Even assuming that broad political and economic stability can be maintained and that the aerospace sector (or key elements of it) does not collapse, it still appears certain that the sector, including the enterprises that support civil space activity, will continue to be severely underfunded, undersupplied, and hard-pressed to retain its skilled personnel. Recently, senior Russian Space Agency (RSA) officials have warned publicly that the Russian human spaceflight program is in imminent jeopardy, although this undoubtedly reflects some degree of posturing for domestic political effect.[18]

The Russian government response could be—as it has been in science and technology more generally—to insist on "maintaining a broad front of research... [forcing] cuts on a random basis, without any rational decisions about what is needed for economic development or military security."[19] To the extent that the Russian authorities are unable or unwilling to establish priorities, Russian enterprises that are key to particular cooperative space projects with the United States will be more-or-less equally at risk across the board.

Alternatively, RSA could decide to set clear priorities among space activities on the basis of their economic or operational value and to suspend support to those that fall too far down the list. Such a decision was made early in the post-Soviet period, when RSA funding was terminated for both the Buran space shuttle and the Energia heavy-lift launch vehicle. Deeper cuts may now be under way, judging by the economic problems currently facing the Russian Mars program and the further delay in the Spectrum-X mission.[20] The U.S. public and private sectors can, of course, influence these decisions over priorities, as they have through procurements for the space station program and joint commercial ventures such as LKE International.

Russian behavior since 1991 apparently reflects both tendencies. Even though the decisions to stop funding the Energia and Buran programs were made at the highest levels of the government,

---

[18] See, e.g., "Manned Space Program in Imminent Jeopardy," *Trud*, Moscow, p. 2, Dec. 10, 1994, in which senior RSA officials warn that the Russian piloted space program "could be terminated in late February 1995" unless more funding is found. The article was published just before the December 15-16, 1994 meeting of the Gore-Chernomyrdin Commission and as the Russian State Duma was debating the 1995 state budget, which suggests a tactical motivation for emphasizing the negative.

[19] Harley Balzer, *Some Thoughts on S&T Cooperation with Russia: Problems of Communication and Perception* (Organisation for Economic Cooperation and Development, in press).

[20] Peter B. deSelding, "Russian Woes Hampering Mars Project," *Space News*, p. 1, Dec. 19-25, 1994.

a year or more passed before Russian officials ceased sending confusing signals in the press about the future of these programs.[21] Similarly, Russian officials delayed postponing the Mars '94 mission until the last possible moment, even though well-documented rumors of the budgetary and technical causes of the delay were circulating a year earlier.[22]

Russian and U.S. program officials could reduce risk from this missed communication to some extent by communicating privately and explicitly with each other about programmatic priorities and funding decisions as they are being made (or as soon as possible thereafter). On occasion, with other partners, such "early warning" has worked extraordinarily well. In 1990-91, for example, NASA's cancellation of the Comet Rendezvous Asteroid Flyby (CRAF) mission, which had significant European (particularly German) involvement, was privately signaled well in advance and has had little lasting political impact. On the other hand, in 1981, the United States gave no warning to the European Space Agency (ESA) before canceling its spacecraft in the International Solar Polar Mission, and European confidence in U.S. reliability as a partner was severely shaken.[23] Frank and open communication with the Russians, although currently more difficult to achieve than such communication with ESA, could prove effective, at least in the non-space-station areas of the relationship.

The problem of "reverse linkage" is more complex and, from the programmatic perspective, may be less tractable than problems of communication. Space cooperation, in general, and space station cooperation, in particular, are highly visible, politically significant components of the overall U.S.-Russian relationship. Vice President Al Gore and Russian Premier Viktor Chernomyrdin are personally engaged, through showcasing space cooperation in their commission's activities. At the same time, the decisions to involve Russia in the space station program, to permit Russian entry into the commercial launch- services market, and to make significant purchases in Russia as part of the new relationship were clearly influenced in large part by the desire to secure continuing Russian adherence to the Missile Technology Control Regime (MTCR). The Gore-Chernomyrdin Commission meeting planned for June 1993 was postponed because of a failure to reach agreement on this issue; at their September 1993 meeting, the two officials announced agreements on MTCR, Russian participation in the space station, and the commitment to spend $400 million on a NASA-RSA contract.[24]

Some observers believe that the linkage between Russian missile-technology-proliferation behavior and space cooperation that has been created in this way could ultimately pose a greater threat to the space station than do technical or programmatic considerations; others believe that the space station relationship is so important to Russia that it provides a strong motivation for continued MTCR compliance.[25] The high profile

---

[21] Part of the confusion may be due to the emergence of space enterprises with some independent ability to keep systems and projects alive on their own. RSC Energia claims that it has continued to maintain and market the Energia launch vehicle (private correspondence from Jeffrey Manber, Managing Director, North American Operations of RSC Energia, to Ray Williamson, OTA, Feb. 3, 1995).

[22] Ibid.

[23] CRAF was paired with the Cassini mission to Saturn, using many of the same spacecraft components and systems that Cassini did, to save money. When it became clear that the cost of the combined program would exceed congressional guidelines, CRAF was canceled while work on Cassini continued. For a discussion of the International Solar Polar Mission (ISPM) cancellation, see U.S. Congress, Office of Technology Assessment, *International Cooperation*, OTA-ISC-239 (Washington, DC: Government Printing Office, July 1985), p. 384.

[24] See Marcia S. Smith, *Space Activities of the United States, CIS, and Other Launching Countries/Organizations: 1957-1993*, 94-347 SPR (Washington, DC: Congressional Research Service, March 1994 (updated periodically)), pp. 36-39.

[25] See, e.g., Marcia S. Smith, "Space Stations," Congressional Research Service Issue Briefs, Washington, DC, October 1994 (updated periodically), pp. 8-9, 16.

afforded the space station in the overall cooperative relationship may also afford it a degree of protection; from this vantage point, the space station may be affected less by negative developments in the overall relationship than are other, lower-profile aspects of space cooperation, including private sector activities.[26]

Businesspeople interviewed by OTA generally find systemic problems in Russia to be a significant brake on developing business relationships. The Russian institutions and legal system, developed under the Soviet regime and undergoing rapid change to fit the new situation, do not yet provide an appropriately stable business environment; observers have described the situation in Russia as resembling that in the United States during the 19th century's "robber baron" era. Sudden, unexplained changes in basic business law and regulations are commonplace, as are corruption and, increasingly, crime. These factors have not deterred U.S. aerospace firms from attempting to establish business relationships in Russia, but they have undoubtedly impeded progress in some cases. The most effective counter to this impediment, most of those interviewed suggest, is to obtain sound specialist advice, expect delays and reverses, and wait out the evolving system.

Relative mutual unfamiliarity, mistrust, and the resulting additional programmatic uncertainty are the inevitable consequence of 30 years of enforced isolation of the two national space programs from one another. For their part, U.S. officials and businesspeople express frustration at their inability to penetrate beyond the largest, best-known of Russian space enterprises; five years ago, they were largely unaware that these enterprises existed. Many Russian managers and officials carry with them entrenched habits of bureaucratic secrecy and tend to resist requests for information, even when those requests have sound business justification and do not jeopardize trade secrets or sensitive technology. Only time and effort on both sides (and, particularly, people in place in each other's establishments) can gradually lower these barriers to the point reached with the United States' traditional cooperative partners; the incorporation of Russian capabilities squarely in the critical path of space station development will necessarily accelerate this process, but at the probable cost of some expensive misunderstandings along the way.[27]

The sheer scale and complexity of the cooperative arrangements with Russia that are in place today for the International Space Station make it unprecedentedly difficult to insulate the program from disruption at any affordable cost. NASA is making a concerted effort to plan for such disruptions, but it acknowledges that a Russian delay or default, depending on when it occurred and what elements of the space station were affected, could cause significant cost or schedule penalties. Moreover, as one observer has suggested, Russian participation may, in fact, be in two critical paths —programmatic and political. Placing Russia in the programmatic critical path means that the program will incur significant delay and resultant increased costs if Russian components are delivered late or not at all. Although very substantial, this risk is at least broadly quantifiable, and from this standpoint, Russian participation is not necessarily essential to the program. The "political critical path" concept addresses whether the United States would be willing to continue the project at all, without Russian involvement, in the current budgetary environment. Those supporting this analysis believe that continuation of the International

---

[26] Russia's unsettled politics make choosing among these hypotheses very difficult.

[27] For example, James T. McKenna, "Mir Docking Device Readied for Rendezvous," *Aviation Week and Space Technology*, p. 72, Sept. 19, 1994, describes difficulties in reaching agreement on the safety certification of the Russian-built docking module for the Shuttle-Mir program. On the other hand, the successful accommodation reached between the two programs, permitting the February 1995 Shuttle-Mir rendezvous to continue despite Russian concerns about a leaking Shuttle thruster, demonstrates what can be accomplished when the stakes are high enough.

Space Station program depends on continued Russian participation.[28]

U.S. officials have focused a great deal of concern on the future viability of the Baikonur launch site, or cosmodrome, which is essential to Russia's participation in the space station, as well as to the commercial use of the Proton launch vehicle for commercial launches. On December 10, 1994, Russian Premier Chernomyrdin and Kazakhstani Prime Minister Akezhan Kazhegeldin signed what appears to be a definitive agreement for the long-term lease of Baikonur to Russia. Earlier, in October 1994, Russian President Boris Yeltsin issued a decree that seemed to resolve internal Russian government differences over the continued maintenance and funding of the cosmodrome. If these measures are implemented and if the resources are made available for restoring the infrastructure at Baikonur and in the supporting city of Leninsk, this concern could recede; first reports are encouraging.[29]

Problems of communication and understanding have their roots both in inherent cultural differences and in the legacy of 75 years of Soviet experience. One participant in OTA's November 1994 workshop declared that "although things are changing very slowly, the most realistic assumption is that the system and attitudes have not changed at all."

U.S. officials and businesspeople emphasized several keys to controlling such risks:

- *Make use of the best available expertise in Russian business law and practices, both to structure relationships correctly and to avoid surprises as much as possible.*
- *Invest in high-quality interpreting and translating.*

- *Never assume a common understanding of terms and concepts; when in doubt, spell them out.*
- *Find out who has the authority to make the needed decisions; many decisions go straight to the top.*
- *Avoid postures or assumptions of superiority. Particularly in technical areas, mutual respect for capabilities and achievements is critical.*

THE ROLE OF GOVERNMENT

This section reviews the roles of government (or, more properly, the U.S. and Russian governments) in civil space cooperation between the United States and Russia. The same observations apply, as well, to cooperation with other states of the former Soviet Union.

■ Governments as Actors

Historically, NASA has resisted "umbrella" space agreements between the United States and other countries and between itself and other space agencies, preferring instead to construct relationships based on a series of individual, self-contained project-level agreements. NASA's rationale for this position is that umbrella agreements tend to create pressures to make projects cooperative whether or not the substantive basis for such projects exists.

This pattern has been broken with the Soviet Union, China, and other countries, including, most recently, Ukraine. In each case, the political symbolism of the umbrella agreement was judged to be such that agency interests were overridden. The current relationship with Russia carries this mutual coupling to a new level of intensity.

---

[28] Kenneth S. Pedersen, Research Professor of International Affairs, Georgetown University, private correspondence with Ray Williamson, OTA, Feb. 13, 1995.

[29] "Working Conditions at Baikonur Improve Following Kazakh Agreement," *Aerospace Daily*, p. 140, Dec. 30, 1994. In late February 1995, a NASA team, returning from work at Baikonur on preparations for launch of the Spektr and Priroda modules, reported that conditions on the spaceport itself were totally satisfactory and that hotel accommodations in Leninsk, except for an absence of hot water, were adequate. NASA also notes that the Russians continue to launch from Baikonur twice each month. On the other hand, one OTA workshop participant questions whether Russia will be able to afford both to maintain the spaceport and to arrest the deterioration of Leninsk.

A second tenet of NASA policy toward international cooperation has been that each side should bring to the venture the financial resources needed to carry out its side of the bargain. The fundamental rationale for this approach is that mutual programmatic interest and priority is best ensured when each party pays its own way and, secondarily, that spending taxpayer dollars abroad is politically risky. Historically, NASA has not opposed international teaming between its contractors and those in cooperating countries; indeed, such teaming has often been needed for the foreign partner to deliver its contribution. Occasionally, as in the case of the space station project, NASA has discouraged its contractors from pursuing such teaming agreements until the governments involved have put the fundamental decisions in place, but the private sector relationships have then followed quickly. Today, for example, U.S. firms and counterparts in Canada, Europe, and Japan have entered into space-station-related contracts and other agreements valued at over $200 million.

Again, the U.S.-Russian space station relationship has broken new ground; in addition to Russian contributions on the usual no-exchange-of-funds basis, direct NASA payments to RSA and directed procurements by NASA contractors from Russian suppliers will total close to $650 million over four years. As discussed in chapter 3, these payments serve important foreign policy goals, although NASA argues that they are also good value and a practical necessity, enabling cooperation to continue during Russia's difficult economic transition.

▮ Governments as Regulators

Historically, U.S. export controls were a highly effective and nearly total block to space trade with the Soviet Union; Russia and the other former Soviet republics remain on the list of proscribed countries in the International Traffic in Arms Regulations (ITAR), meaning that the Secretary of State (or his designee) must grant a waiver before any export of goods covered by those regulations can take place.

In 1993, partly in recognition of the end of the Cold War, the United States revised the ITAR Munitions List, placing almost all civil space hardware (except for launch vehicles and associated technology, remote-sensing satellites, and communications satellites and components with significant military utility) under the control of the Department of Commerce. Significantly, however, detailed design and manufacturing information on all space hardware and software remains on the Munitions List.

NASA has negotiated with the Departments of State, Defense, and Commerce a blanket data-export authorization for the space station project, which permits the export of all interface and specification data necessary for Russia to carry out its responsibilities, on the same basis that such data are exported to the other partners. Other cooperative activities, such as the export of instruments and related data for flight on Russian spacecraft, continue to require case-by-case authorization.

Private sector activities are still subject to ITAR in most cases because, almost without exception, the first stage of developing a joint venture or other cooperative relationship involves an "export" of technical data for the purpose of initiating substantive discussions. During the OTA workshop, several participants from the private sector complained that the process continues to place an onerous burden on their activities, often including a requirement that their negotiations be monitored by Defense Department personnel. Others noted that the U.S. government uses the licensing process to pursue its policy goals in areas such as space-launch trade and missile-technology proliferation, holding back on license approvals until appropriate agreements are obtained, as in the case of Boeing's proposed joint venture to market a Ukrainian launch vehicle's services. Others commented that in many cases, the problem appeared to be less the substance of the regulations themselves than the "old Cold Warrior" attitudes they ascribe to the officials and military officers involved.

Although information is more fragmentary and Russian institutions in the field of technology-transfer control are less well-developed than they are in the United States, there have been some indications of impediments to expanded cooperation at work in Russia, too. Complaints about Russia's selling off its technical birthright for pennies on the dollar have been common in the Russian press. One firm reported that an important deal was being delayed because of lack of approval for transfer of the technology involved by a Russian interagency group concerned with technology security. Because of extensive commonality between Russian remote-sensing systems and their military counterparts, security concerns have imposed considerable overhead on efforts to market remote-sensing data products in the West, some businesspeople report.

Another important regulatory area is the field of space-launch trade. As mentioned above, one of the most important motivations for Russian agreement to abide by the MTCR was U.S. willingness to allow Russian space-launch services to compete to launch U.S.-built commercial satellites.

Competitive issues aside, potential earnings from commercial launch sales may be important to keeping Russian rocket designers employed at home rather than offering their services to Third World missile programs. The current agreement, signed in September 1993, is designed to be transitional and allows Russia a total of only eight geostationary orbit launches through the year 2000. However, by the end of 1994, LKE International had reportedly won 15 firm contracts or options worth more than $1 billion and was expected to fill the Russian quota with firm launch contracts during 1995.[30] The Clinton Administration is coming under pressure from Lockheed, U.S. satellite manufacturers, and the Russians to expand the quota, particularly in light of the conclusion in January 1995 of a much more liberal agreement between the United States and China. Meanwhile, U.S. launch-vehicle manufacturers and Europe's Arianespace complain that the current agreement's price provisions, in particular, are not being adequately enforced, and those companies oppose any further market share for Russia.[31]

---

[30] "Lockheed Signs Up 15 Launches for Proton Venture," *Aerospace Daily*, p. 390, Dec. 20, 1994. Only very limited information on the financial arrangements between the partners is publicly available, but Lockheed's investment to date has apparently been modest compared with the potential revenues involved.

[31] Andrew Lawler, "U.S. To Begin Launch Talks with China, Russia," *Space News*, p. 1, 20, Sept. 12-28, 1994.

Domestic
Impact
and
Proliferation
Concerns | 6

INTRODUCTION

Some observers express concern that U.S.-Russian commercial cooperation might cost U.S. aerospace jobs, erode the country's space-technology base, and undercut competitiveness of U.S. companies by transferring sophisticated technology to a foreign competitor. In fact, the costs of cooperation will have to be balanced against the potential benefits, some of which may extend well beyond any specific project. For example, U.S. officials are deeply concerned about the proliferation of ballistic-missile technologies to developing countries. Russia is a potential source of missiles, components, and expertise, whose transfer could benefit a country trying to develop its own ballistic-missile capability.

A combination of economic incentives and economic sanctions might be effective in curtailing the sale of hardware useful in the development and deployment of ballistic missiles, and it might help to keep the rocket scientists, whose expertise is an essential part of a working ballistic-missile program, from leaving Russia to work for a developing nation that would pay well for their services. A collapsing aerospace industry, with massive layoffs, dwindling salaries, and no jobs for young scientists and engineers who are just starting out, puts great pressure on employees to seek greener pastures outside Russia. Of particular concern are those scientists who would aid states, such as Iran, that are actively hostile to the United States. Although emigration restrictions seem to have been effective in preventing some at-

tempts at expatriation by aerospace engineers,[1] one long-term solution to the "brain drain" problem is a stable, viable Russian aerospace industry.

This chapter summarizes some of the issues that come into play in a consideration of future U.S.-Russian cooperation.

DOMESTIC IMPACT

The effect on the U.S. aerospace industry of Russia's entry into the international space-launch market will depend on how the United States decides to structure commercial cooperation with the Russians and on which part of the industry attention is focused. On the one hand, access to different and up-to-date technologies, production and processing methods, and cheaper hardware could make the U.S. aerospace industry stronger in an ever more competitive world market for space-related services. On the other hand, cooperative arrangements could also lead to unwanted technology transfer, strengthening of a competitor, loss of domestic production jobs, and a weakening of U.S. capabilities because of dependence on a foreign source.

The United States is in the process of deciding how to evolve its space technology so that it can be as efficient as possible in meeting the domestic need for access to space and in competing in the international space-launch-services market.[2] Because the requirements of the Soviet/Russian space program have historically been different from those of the U.S. program, Russia has developed systems with different operational and design characteristics. Access to Russian technological innovations could offer U.S. decisionmakers a wider range of design possibilities from which to choose, some of which have already been tested and implemented by the Russians. Some elements of their aerospace industry that might enhance U.S. capabilities are automated launch capabilities, less expensive hardware, advanced materials and materials processing, computational methods, and technical expertise.[3]

∎ U.S. Job Market and Industrial Base

The current U.S.-Russian agreement on international trade in commercial space-launch services seeks to prevent Russia from providing space-launch services at prices more than 7.5 percent below "the lowest bid or offer by a commercial space launch service provider from a market economy country."[4] It also limits the number of principal-payload[5] launches that the Russians can sell on the international market to eight[6] between now and the year 2000. Both of these quantitative limits reflect an attempt to protect domestic providers of medium- to heavy-lift launch services from encountering unfair competition from the Russian

---

[1] In December 1992, more than 50 Russian rocket scientists were stopped at Moscow's Sheremetyevo Airport. They had been recruited by North Korea with the promise of salaries much higher than they could command in Russia, according to one report (U.S. Congress, Office of Technology Assessment, *Proliferation and the Former Soviet Union,* OTA-ISS-605 (Washington, DC: U.S. Government Printing Office, September 1994), pp. 32-33, 643). The report goes on to point out, "In spite of the fact that the arrest has a positive aspect, reinforcing the belief that the Russian authorities are alert to foreign efforts to recruit or corrupt their specialists, there is also a negative aspect: the event demonstrates an active, advanced effort by a state to gain technologies controlled by an international nonproliferation regime."

[2] For a discussion of the objectives and possible effects of the Clinton Administration's *Nation Space Transportation Policy,* see U.S. Congress, Office of Technology Assessment, *The National Space Transportation Policy: Issues for Congress* (Washington, DC: U.S. Government Printing Office), forthcoming, spring 1995.

[3] Chapter 5 presents a more detailed catalogue of Russian capabilities that could be useful to the U.S. aerospace industry.

[4] "Agreement Between the Government of the United States of America and the Government of the Russian Federation Regarding International Trade in Commercial Space Launch Services," 1993, p. 8.

[5] A principal payload is a telecommunications satellite or, in the absence of a telecommunications satellite, any other spacecraft or combination of spacecraft.

[6] This does not include the scheduled launch of an INMARSAT 3 satellite on a Russian Proton booster. The payloads referred to are commercial payloads; no limit is placed on the number of payloads that can be launched with either the Russian or U.S. government as the customer.

aerospace industry, which is heavily subsidized by the Russian government, from the top-level manufacturer down through all lower-tier suppliers. There is also excess capacity in the Russian aerospace industry, dormant now, that could presumably be brought into play if sufficient demand develops. The overall effect is that the Russian aerospace industry, if not constrained, might be able to meet a large demand for launch services at prices much lower than U.S. firms could offer.

The U.S. aerospace industry is made up of different segments with differing needs, which complicates the attempt to predict the effect on jobs of using Russian launch services. Removal of all quotas on the number and price of Russian launches might be burdensome competition to a U.S. launch-service provider and, at the same time, a boon to a provider of on-orbit capabilities who must pay to launch its satellites to orbit. Whether such a tradeoff would result in a net increase or decrease of jobs in the aerospace industry as a whole is not clear. Even a **net** increase in jobs might be small consolation to a launch-service provider that loses out. Some observers argue that Russian entry into the launch-vehicle market might result in an increase of business in the aerospace industry as a whole because of Russian technological capabilities that make launch services cheaper. In that case, having Russian hardware and technical expertise available to U.S. industry for marketing at home and abroad could position the U.S. aerospace industry to capture a larger share of the expanded overall market, even while it is losing market share in the launch-services component of this market.

It might also be possible for domestic firms to take advantage of Russian launch capabilities directly. As an example, the formation of Lockheed-Krunichev-Energia (LKE) International is an attempt by Lockheed to market Proton launches to geosynchronous orbit. LKE International argues that it will not be taking market share away from the U.S. Atlas or Titan, but from the French Ariane 4 and 5. Representatives of the U.S. launch industry at an Office of Technology Assessment workshop, "Lower Industrial Tiers of the Space Launch Vehicle Industry," held in March 1995 expressed another viewpoint: the domestic launch industry is struggling and does not need another competitor in the medium-to-heavy launch-service market, irrespective of any possible enhancement of U.S. capabilities through cooperation with the Russians.[7]

Apparently, the effect that any given cooperative venture with the Russians will have on jobs in the U.S. space industry will depend on how that cooperation is structured. Several possible arrangements are:

- *Independent contribution.* Have each side design and develop its contribution separately and provide the other side with interface documents only. This type of arrangement has the advantage of making it possible to control technology transfer between the parties involved. But the components of a joint venture provided by a foreign entity are not manufactured in the United States, so there would be no contribution to U.S. manufacturing jobs. A joint arrangement with independent contributions from both parties could, however, provide a new service, or an existing service at a lower price, thereby benefiting the U.S.-based partner.

- *Commercial buy.* In this case, a propulsion firm such as Pratt and Whitney or Aerojet might buy Russian rocket engines that could be made compatible with U.S. boosters. Although such a buy will probably lose jobs for the engine-manufacturing segment of the domestic industry, in most cases, testing and systems engineering will still be required. Also, cheaper engines might make U.S. launch services more competitive, potentially increasing business and creating jobs in that sector of the industry

---

[7] U.S. Congress, Office of Technology Assessment, *The National Space Transportation Policy: Issues for Congress* (Washington, DC: U.S. Government Printing Office), forthcoming, spring 1995.

and in others stimulated by low-cost launch services.

- *Licensing technology.* A U.S. firm could buy a license for a given engine technology and set up its own production line. This licensing of technology would result in increased employment for U.S. workers if it is successful in producing a product. It could also make those parts of the industry that depend on the product of the licensed technology more competitive in the world market.

- *Joint development.* In a joint business venture that seeks to develop a new service, the venture can benefit from the technological expertise that each side brings with it. Such ventures could bring technological advancement to both sides, which might then create new markets for the products that would result from cooperation.

The United States must also decide how much of its industrial base should be maintained to meet national security needs and to ensure access to space. Making use of existing Russian technology could reduce the amount of research and development required of U.S. firms, resulting in reduced costs, but it could also undercut the development of U.S. capabilities in certain areas. Because the space industry is considered to be indispensable to the security of the United States, many argue that the United States should develop and maintain its own capabilities in certain critical areas to prevent any weakening in its own technological base. In line with that reasoning is the National Space Transportation Policy, which states that the U.S.

government will not purchase launches on vehicles not manufactured in the United States.[8] The Department of Defense (DOD) is willing to use launch systems that have foreign components and technology, but only in such a way that foreign suppliers cannot deny DOD access to space.[9] Although this might result in higher costs to the government, it ensures that the United States will be able to fulfill its space-related national security needs without depending on foreign suppliers of launch services.

■ Technology Transfer

Cooperative ventures entail the risk of transfer of domestic technologies that could be used to strengthen a competitor's position in the international aerospace market. Policymakers disagree over how effective specific means to prevent such transfer can really be, but present policy is clearly in the direction of loosening trade restrictions. Specifically, many items having to do with satellites and satellite technology have been moved from the U.S. Munitions List[10] onto the Commerce Control List, effectively making it easier to trade in those items.[11] There are recent reports that further loosening of restrictions is being worked out between the Department of State and the Department of Commerce.[12]

PROLIFERATION CONCERNS[13]

The principal current attempt to limit proliferation of long-range delivery systems capable of delivering weapons of mass destruction (nuclear, chemical, and biological weapons) is the Missile

---

[8] The White House, Office of Science and Technology Policy, *Fact Sheet: National Space Transportation Policy,* Aug. 5, 1994, section VI.

[9] *DOD Implementation Plan for National Space Transportation Policy,* PDD/NSTC-4, Nov. 4, 1994.

[10] Code of Federal Regulations 22, ch. 1, subch. M—International Traffic in Arms Regulations, Part 121—The United States Munitions List, 1994, pp. 383-402.

[11] The U.S. Munitions List regulates export of items considered to have explicit military value. Those exports are regulated by the State Department under the Arms Export Control Act (P. L. 90-629). The Commerce Control List includes dual-use items that have both civil and military application. Those items are controlled by the Commerce Department under the Export Administration Act (P. L. 96-72).

[12] Warren Ferster, "Satellite Export Controls To Ease," *Space News,* p. 1, Feb. 20-26, 1995.

[13] Most of the material in this section is taken from chapter 5 of U.S. Congress, Office of Technology Assessment, *Technologies Underlying Weapons of Mass Destruction,* OTA-BP-ISC-115 (Washington, DC: U.S. Government Printing Office, December 1993).

Technology Control Regime (MTCR), created in 1987 by the United States and other Western industrialized nations. The MTCR established a presumption to deny the transfer of ballistic missiles with ranges greater than 300 kilometers and payload capacities greater than 500 kilograms to nonmember nations. These guidelines have since been extended to cover any systems "intended to deliver weapons of mass destruction." Russia has pledged to join the MTCR and has agreed to abide by its rules until it becomes a full-fledged member. Participation in the MTCR requires that Russia prohibit the transfer of complete systems, components that could be used to make complete systems, and technology involved in the production of components or of complete systems.

Missile systems and space-launch systems have much in common, and arguments arise over whether a particular technology is best suited to one type of system or the other, or could be used for both. Despite having many components and technologies in common, space-launch systems differ from vehicles designed to reenter the Earth's atmosphere and strike targets on the ground. Space-launch systems do not require the sophisticated guidance needed for long-range ballistic missiles; a 10-kilometer error is tolerable for putting a payload into orbit, but is a great tactical impediment when trying to hit a long-range target, even for weapons of mass destruction. There are many other technological barriers that separate space-launch systems from working ballistic-missile systems, including the need for sophisticated materials-processing capabilities and advanced guidance systems. Despite all the technological difficulties involved in producing a working ballistic-missile system, testing and development of weapon-delivery systems can be accomplished under the guise of developing a space-launch program. The prudent assumption is that any country

that has space-launch vehicles should be considered capable of developing ballistic missiles.

Economic shortfalls in the space sector and throughout the Russian economy make the sale of expensive, high-technology missile components and systems extremely attractive. In 1992, India contracted with Russia to buy a liquid-oxygen/liquid-hydrogen engine to be used as the upper stage for its Geosynchronous Satellite Launch Vehicle (GSLV). Both India and Russia resisted attempts by the United States to declare the deal to be a violation of the MTCR, which would have triggered sanctions that U.S. law requires be applied against states engaged in such transfer. Finally, in 1993, and against the wishes of the Indian government, Russia agreed to break its contract with India and withhold the engine technology.[14]

The question remains of what the United States can do to forestall the proliferation of technology, components, and expertise from Russia to developing nations. Even if Russia is willing to abide by the MTCR, as it has pledged to do, and prohibit the export of hardware useful in ballistic missiles, it might not be able to prevent the emigration of rocket scientists seeking to escape stifling economic conditions that are aggravated by the present state of the Russian space program. Despite Russia's apparent concern over the loss of its aerospace engineers, it might not be able to prevent their departure in all cases. People with expertise can freely emigrate from Russia to neighboring countries in the Newly Independent States (NIS), and keeping track of where they go from there might not be possible.

The United States might consider it in the interest of global nonproliferation to try to ensure that the Russian space program has the greatest possible chance of remaining healthy and capable of

---

[14] Four of the engines were sold to India by Russia. The United States' main concern was the potential military uses of the technology that was being transferred rather than the sale of the cryogenic engines themselves. Observers differed in their opinions about the usefulness of cryogenic engines for weapons systems. Weapons systems require constant readiness, and cryogenic engines take a long time to prepare for launch. There is no question, however, that some of the technology involved in the transfer could be beneficial to the development of long-range ballistic missiles.

retaining its experts. A similar kind of decision arises in the case of the proliferation of nuclear-weapons expertise, or brain drain. Attempting to prevent the proliferation of nuclear weapons is probably more difficult because the scale of the operation required to build some kinds of nuclear weapon is small (particularly if the required nuclear material—enriched uranium or plutonium—is available on the black market), while a ballistic-missile program requires the integration of a variety of complex and sometimes large systems. Nonetheless, the U.S. government's response to the brain drain in the area of nuclear-weapons technology was to provide some direct funding to scientific researchers responsible for the development and engineering of nuclear, chemical, and biological weapons in an effort to keep them employed in areas other than the development of those weapons.[15]

Many of the scientists and engineers in the Russian civil and military space programs have expertise that could be usefully applied to space science missions. Even during Cold War periods when the political atmosphere made larger, high-profile cooperative science efforts unacceptable, small, low-profile science projects involving Russian and U.S. scientists continued. That ongoing cooperation kept the lines of communication between the two countries open and fostered commonality of interest. With the lessening of tensions following the end of the Cold War, opportunities have increased for including Russia in international science projects and for joint U.S.-Russian science missions.

---

[15] Since FY1992, the Nunn-Lugar amendment to Public Law 102-228 and subsequent legislation have authorized the transfer of $1.6 billion of Department of Defense funds to help accomplish the destruction and secure storage of weapons of mass destruction. Of that money, $25 million was to be the 1994 U.S. contribution to the International Science and Technology Center (ISTC), which would provide research opportunities for former Soviet Union scientists in collaborative efforts with Western scientists. See, U.S. Congress, Office of Technology Assessment, *Proliferation and the Former Soviet Union,* OTA-ISS-605 (Washington, DC: U.S. Government Printing Office, September 1994), pp. 23-28. Some U.S. private foundations have also made money available to Russian research institutions to try to curtail the proliferation of nuclear-weapons expertise.

Appendix A:
1992 and Subsequent
U.S.-U.S.S.R.
Space Agreements A

APPENDIX A1:

Agreement Between the United States of America and the Russian Federation Concerning Cooperation in the Exploration and Use of Outer Space for Peaceful Purposes (June 1992)

The United States of America and the Russian Federation, hereinafter referred to as the Parties;

Considering the role of the two states in the exploration and use of outer space for peaceful purposes;

Desiring to make the results of the exploration and use of outer space available for the benefit of the peoples of the two states and of all peoples of the world;

Considering the respective interest of the Parties in the potential for commercial applications of space technologies for the general benefit;

Taking into consideration the provisions of the Treaty on Principles Governing the Activities of States in the Exploration and Use of Outer Space, including the Moon and other Celestial Bodies, and other multilateral agreements regarding the exploration and use of outer space to which both states are Parties;

Expressing their satisfaction with cooperative accomplishments in the fields of astronomy and astrophysics, earth sciences, space biology and medicine, solar system exploration and solar terrestrial physics, as well as their desire to continue and enhance cooperation in these and other fields;

Have agreed as follows:

■ Article I

The Parties, through their implementing agencies, shall carry out civil space cooperation in the fields of space science, space exploration, space applications and space technology on the basis of equality, reciprocity and mutual benefit.

Cooperation may include human and robotic space flight projects, ground-based operations and experiments and other activities in such areas as:

—Monitoring the global environment from space;

—Space Shuttle and Mir Space Station missions involving the participation of U.S. astronauts and Russian cosmonauts;

—Safety of space flight activities;

—Space biology and medicine; and,

—Examining the possibilities of working together in other areas, such as the exploration of Mars.

■ Article II

For purposes of developing and carrying out the cooperation envisaged in Article I of this Agreement, the Parties hereby designate, respectively, as their principal implementing agencies the National Aeronautics and Space Administration for the United States and the Russian Space Agency for the Russian Federation.

The Parties may designate additional implementing agencies as they deem necessary to facilitate the conduct of specific cooperative activities in the fields enumerated in Article I of this Agreement.

Each of the cooperative projects may be the subject of a specific written agreement between the designated implementing agencies that defines the nature and scope of the project, the individual and joint responsibilities of the designated implementing agencies related to the project, financial arrangements, if any, and the protection of intellectual property consistent with the provisions of this Agreement.

■ Article III

Cooperative activities under this Agreement shall be conducted in accordance with national laws and regulations of each party, and shall be within the limits of available funds.

■ Article IV

The Parties shall hold annual consultations on civil space cooperation in order to provide a mechanism for government-level review of ongoing bilateral cooperation under this Agreement and to exchange views on such various space matters. These consultations could also provide the principal means for presenting proposals for new activities falling within the scope of this Agreement.

■ Article V

This Agreement shall be without prejudice to the cooperation of either Party with other states and international organizations.

■ Article VI

The Parties shall ensure adequate and effective protection of intellectual property created or furnished under this Agreement and relevant agreements concluded pursuant to Article II of this Agreement. Where allocation of rights to intellectual property is provided for in such agreements, the allocation shall be made in accordance with the Annex attached hereto which is an integral part of this Agreement. To the extent that it is necessary and appropriate, such agreements may contain different provisions for protection and allocation of intellectual property.

■ Article VII

This Agreement shall enter into force upon signature by the Parties and shall remain in force for five years. It may be extended for further five-year periods by an exchange of diplomatic notes. This Agreement may be terminated by either Party on six months written notice, through the diplomatic channel, to the other Party.

DONE at Washington, in duplicate, this seventeenth day of June, 1992, in the English and Russian languages, both texts being equally authentic.

FOR THE UNITED STATES OF AMERICA: FOR THE RUSSIAN FEDERATION:

George Bush   Boris Yeltsin

ANNEX: INTELLECTUAL PROPERTY

Pursuant to Article VI of this Agreement:

The Parties shall ensure adequate and effective protection of intellectual property created or furnished under this Agreement and relevant agreements concluded pursuant to Article II of this Agreement. The Parties agree to notify one another in a timely fashion of any inventions or copyrighted works arising under this Agreement and to seek protection for such intellectual property in a timely fashion. Rights to such intellectual property shall be allocated as provided in this Annex.

▮ I. Scope

a. This annex is applicable to all cooperative activities undertaken pursuant to this Agreement, except as otherwise specifically agreed by the Parties or their designees.

b. For purposes of this Agreement, "intellectual property" shall have the meaning found in Article 2 of the convention establishing the World Intellectual Property Organization, done at Stockholm, July 14, 1967.

c. This Annex addresses the allocation of rights, interests, and royalties between the Parties. Each Party shall ensure that the other Party can obtain the rights to intellectual property allocated in accordance with the Annex, by obtaining those rights from its own participants through contracts or other legal means, if necessary. This Annex does not otherwise alter or prejudice the allocation between a Party and its participants, which shall be determined by that Party's laws and practices.

d. Disputes concerning intellectual property arising under this Agreement should be resolved through discussions between the concerned participating institutions or, if necessary, the Parties or their designees. Upon mutual agreement of the Parties, a dispute shall be submitted to an arbitral tribunal for binding arbitration in accordance with the applicable rules of international law. Unless the Parties or their designees agree otherwise in writing, the arbitration rules of UNCITRAL shall govern.

e. Termination or expiration of this Agreement shall not affect rights or obligations under this Annex.

▮ II. Allocation of Rights

a. Each party shall be entitled to a non-exclusive, irrevocable, royalty-free license in all countries to translate, reproduce, and publicly distribute scientific and technical journal Articles, reports, and books directly arising from cooperation under this Agreement. All publicly distributed copies of a copyrighted work prepared under this provision shall indicate the names of the authors of the work unless an author explicitly declines to be named.

b. Rights to all forms of intellectual property, other than those rights described in Section II(a) above, shall be allocated as follows:

1. Visiting researchers and scientists visiting primarily in furtherance of their education shall receive intellectual property rights under the policies of the host institution. In addition, each visiting researcher or scientist named as an inventor shall be entitled to share in a portion of any royalties earned by the host institution from the licensing of such intellectual property.

2. (a) For intellectual property created during joint research with participation from the two Parties, for example, when the Parties, participating institutions, or participating personnel have agreed in advance on the scope of work, each Party shall be entitled to obtain all rights and interests in its own country. Rights and interests in third countries will be determined in agreements concluded pursuant to Article II of this Agreement. The rights to intellectual property shall be allocated with due regard for the economic, scientific and technological contributions from each Party to the creation of intellectual property. If research is not designated as "joint research" in the relevant agreement concluded pursuant to Article II of this Agreement, rights to intel-

lectual property arising from the research shall be allocated in accordance with Paragraph IIb1. In addition, each person named as an inventor shall be entitled to share in a portion of any royalties earned by their institution from the licensing of the property.

(b) Notwithstanding Paragraph IIb2(a), if a type of intellectual property is available under the laws of one Party but not the other Party, the Party whose laws provide for this type of protection shall be entitled to all rights and interests in all countries which provide rights to such intellectual property. Persons named as inventors of the property shall nonetheless be entitled to royalties as provided in Paragraph IIb2(a).

■ III. Business-Confidential Information

In the event that information identified in a timely fashion as business-confidential is furnished or created under the Agreement, each Party and its participants shall protect such information in accordance with applicable laws, regulations, and administrative practice. Information may be identified as "business-confidential" if a person having the information may derive an economic benefit from it or may obtain a competitive advantage over those who do not have it, the information is not generally known or publicly available from other sources, and the owner has not previously made the information available without imposing in a timely manner an obligation to keep it confidential.

APPENDIX A2:

Protocol to the Implementing Agreement Between the National Aeronautics and Space Administration of the United States of America and the Russian Space Agency of the Russian Federation on Human Space Flight Cooperation of October 5, 1992

∎ Preamble

The National Aeronautics and Space Administration (hereafter referred to as "NASA"), and the Russian Space Agency (hereafter referred to as "RSA"), jointly referred to as "the Parties;"

Consistent with the Joint Statement on Cooperation in Space issued by Vice President Gore and Prime Minister Chernomyrdin on September 2, 1993; desiring to broaden the scope of the Implementing Agreement of October 5, 1992, on Human Space Flight Cooperation (hereinafter the October 5, 1992 Agreement) to encompass an expanded program of activities for cooperation involving the Russian Mir-1 Space Station and the U.S. Space Shuttle Program;

Having decided that the enhanced cooperative program will consist of a number of inter-related projects in two phases;

Having determined that Phase One will include those activities described in the October 5, 1992, Agreement and known as the Shuttle-Mir Program, including the exchange of the Russian Mir-1 crew and crew member participation in joint mission science, as well as additional astronaut flights, Space Shuttle dockings with Mir-1, and other activities;

Having further determined that Phase Two of the enhanced cooperative program will involve use of a Russian Mir module of the next generation mated with a U.S. laboratory module operated on a human-tended basis in conjunction with the Space Shuttle, operating in a 51.6 degree orbit which is accessible by both U.S. and Russian resources, to perform precursor activities for future space station-related activities of each Party, with launch to occur in 1997; and

Intending that activities in Phase Two would be effected through subsequent specific agreement(s) between the Parties.

Have agreed as follows:

∎ Article I: Description of Additional Activities

1. This Protocol forms an integral part of the October 5, 1992 Agreement.
2. An additional Russian cosmonaut flight on the Space Shuttle will take place in 1995. The back-up cosmonaut currently in training at NASA's Johnson Space Center will be the primary cosmonaut for that flight, with the STS-60 primary cosmonaut acting as back-up. During this mission, the Shuttle will perform a rendezvous with the Mir-1 Space Station and will approach to a safe distance, as determined by the Flight Operations and Systems Integration Joint Working Group established pursuant to the October 5, 1992 Agreement.
3. The Space Shuttle will rendezvous and dock with Mir-1 in October-November 1995, and, if necessary, the crew will include Russian cosmonauts. Mir-1 equipment, including power supply and life support system elements, will also be carried. The crew will return on the same Space Shuttle mission. This mission will include activities on Mir-1 and possible extravehicular activities to upgrade solar arrays. The extravehicular activities may involve astronauts of other international partners of the Parties.

4. NASA-designated astronauts will fly on the Mir-1 space station for an additional 21 months for a Phase One total of two years. This will include at least four astronaut flights. Additional flights will be by mutual agreement.

5. The Space Shuttle will dock with Mir-1 up to ten times. The Shuttle flights will be used for crew exchange, technological experiments, logistics or sample return. Some of those flights will be dedicated to resources and equipment necessary for life extension of Mir-1. For schedule adjustments of less than two weeks, both sides agree to attempt to accommodate such adjustments without impacting the overall schedule of flights. Schedule adjustments of greater than two weeks will be resolved on a case-by-case basis through consultations between NASA and RSA.

6. A specific program of technological and scientific research, including utilization of the Mir-1 Spekter and Priroda modules, equipped with U.S. experiments, to undertake a wide-scale research program, will be developed by the Mission Science Joint Working Group established pursuant to the October 5, 1992 Agreement. The activities carried out in this program will expand ongoing research in biotechnology, materials sciences, biomedical sciences, Earth observations and technology.

7. Technology and engineering demonstrations applicable to future space station activities will be defined. Potential areas include but are not limited to: automated rendezvous and docking, electrical power systems, life support, command and control, microgravity isolation system, and data management and collection. Joint crew operations will be examined as well.

8. The Parties consider it reasonable to initiate in 1993 the joint development of a solar dynamic power system with a test flight on the Space Shuttle and Mir in 1996, the joint development of spacecraft environmental control and life support systems, and the joint development of a common space suit.

9. The Parties will initiate a joint crew medical support program for the benefit of both sides' crew members, including the development of common standards, requirements, procedures, databases, and countermeasures. Supporting ground systems may also be jointly operated, including telemedicine links and other activities.

10. The Space Shuttle will support the above activities, including launch and return transportation of hardware, material, and crew members. The Shuttle may also support extravehicular and other space activities.

11. Consistent with U.S. law, and subject to the availability of appropriated funds, NASA will provide both compensation to the RSA for services to be provided during Phase One in the amount of US $100 million in FY 1994, and additional funding of US $300 million for compensation of Phase One and for mutually-agreed upon Phase Two activities will be provided through 1997. This funding will take place through subsequent NASA-RSA and/or through industry-to-industry arrangements. Reimbursable activities covered by the above arrangements and described in paragraphs 3-8 will proceed after these arrangements are in place and after this Protocol enter into force in accordance with Article III. Specific Phase One activities, schedules and financial plans will be included in separate documents.

12. Implementation decisions on each part of this program will be based on the cost of each part of the program, relative benefits to each Party, and relationship to future space station activities of the Parties.

13. The additional activities will not interfere with or otherwise affect any existing, independent obligations either Party may have to other international partners.

■ Article II: Joint Implementation Teams

The coordination and implementation of the activities described herein will be conducted through the Joint Working Groups established pursuant to the October 5, 1992 Agreement or such other joint bodies as may be established by mutual agreement.

■ Article III: Entry into Force

This Protocol will enter into force upon an exchange of diplomatic notes between the Governments of the United States of America and the Russian Federation confirming acceptance of its terms and that all necessary legal requirements for entry into force have been fulfilled.

IN WITNESS WHEREOF the undersigned, being duly authorized by their respective Governments, have signed this Protocol.

Done at Moscow, in duplicate, this sixteenth day of December, 1993, in the English and Russian languages, both texts being equally authentic.

FOR THE NATIONAL AERONAUTICS AND
SPACE ADMINISTRATION OF THE
UNITED STATES OF AMERICA:

FOR THE RUSSIAN SPACE AGENCY OF
THE RUSSIAN FEDERATION:

Dan Goldin

Yuri Koptev

APPENDIX A3:

Interim Agreement Between the National Aeronautics and Space Administration of the United States of America and the Russian Space Agency for the Conduct of Activities Leading to Russian Partnership in the Detailed Design, Development, Operation and Utilization of the Permanently Manned Civil Space Station

The National Aeronautics and Space Administration of the United States of America (hereinafter referred to as "NASA"), and the Russian Space Agency (hereinafter referred to as "RSA"), hereinafter also referred to as the "Parties",

RECOGNIZING the Agreement between the United States of America and the Russian Federation Concerning Cooperation in the Exploration and Use of Outer Space for Peaceful Purposes of June 17, 1992;

RECOGNIZING the successful cooperation being conducted by NASA and RSA under the Human Space Flight Agreement of October 5, 1992, and the Protocol to that Agreement of December 16, 1993;

RECALLING the Summit Meeting of April 3, 1993, between Presidents Clinton and Yeltsin which established the U.S.-Russian Joint Commission on Energy and Space;

RECALLING the Joint Statement of September 2, 1993, on Cooperation in Space issued by the U.S.-Russian Joint Commission on Energy and Space chaired by Vice President Gore and Prime Minister Chernomyrdin;

RECALLING the Joint Statement of December 16, 1993, on Space Cooperation issued by the U.S.-Russian Joint Commission on Economic and Technological Cooperation chaired by Vice President Gore and Prime Minister Chernomyrdin;

RECOGNIZING the Joint Invitation at the Occasion of the Intergovernmental Meeting of the Space Station Partners in Washington, DC, on December 6, 1993; and further recognizing the acceptance of the invitation by the Government of the Russian Federation on December 17, 1993;

NOTING the obligations of the United States of America and NASA pursuant to:

The Agreement Among the Government of the United States of America, Governments of Member States of the European Space Agency, the Government of Japan, and the Government of Canada on Cooperation in the Detailed Design, Development, Operation, and Utilization of the Permanently Manned Civil Space Station of September 29, 1988 (the Agreement is referred to hereinafter as the "IGA"; the Governments of the United States, Japan, Canada, and the European Governments collectively, are hereinafter referred to as the "Partners");

Memoranda of understanding on cooperation in the detailed design, development, operation, and utilization of the permanently manned civil space station between NASA and:

The Ministry of State for Science and Technology of Canada (September 29, 1988), and further noting that upon its establishment on March 1, 1989, the Canadian Space Agency (CSA) assumed responsibility for the execution of the Canadian Space Station Program from MOSST;

The European Space Agency (ESA) (September 29, 1988); and

The Government of Japan (March 14, 1989), and further noting the designation by the Government of Japan of the Science and Technology Agency (STA) as its cooperating agency; (CSA, ESA, and STA are hereinafter collectively referred to as the "Cooperating Agencies of the Space Station Partners");

NOTING the commitments of NASA in the Space Station Program Implementation Plan of September 7, 1993;

RECOGNIZING the Addendum to the Space Station Program Implementation Plan of November 2, 1993, hereinafter referred to as the "Addendum"; and

RECOGNIZING the Joint Statement on Negotiations Related to the Integration of Russia into the Space Station Partnership issued by the Partners and the Government of the Russian Federation at the Intergovernmental Meeting on March 18, 1994, and the adoption of the following papers: "Changes in the Legal Framework to Include Russia as a Partner" and "Modalities for Forthcoming Negotiations on the Space Station Agreements;"

HAVE AGREED AS FOLLOWS:

∎ Article 1—Objectives

1.1 This Agreement sets out the terms and conditions for NASA and RSA cooperation in activities related to the initial participation in the Space Station Program by organizations or entities of, or related to, the Government of the Russian Federation. This cooperation, an integrated partnership among NASA, RSA and the Cooperating Agencies of the Space Station Partners, will contribute significantly to achieving the goal of a single, integrated international Space Station that will enhance the use of space for the benefit of all participating nations and humanity. The Parties are also cooperating in additional activities pursuant to other agreements, including precursor activities to the cooperation on the Space Station described in this Agreement. These precursor activities, referred to as Phase 1, involve efforts to achieve significant risk reduction in the overall program which are not subject to this Agreement. This Agreement covers later phases of the cooperation: the detailed design, development, operation and utilization of the Space Station.

1.2 This Agreement provides for:

Initial cooperation between NASA and RSA to integrate RSA into the planning process for detailed design, development, operation and utilization of the Space Station pending completion of programmatic steps and entry into force of legal arrangements for a redesigned Space Station with integral Russian participation;

Descriptions of managerial, technical and operational interfaces which are necessary to ensure effective coordination and compatibility between Parties' activities; and

Establishment of specified legal obligations, in connection with Russian participation in the Space Station Program.

1.3 In particular, the purpose of this Agreement is to integrate RSA, to the maximum extent possible, into Space Station management mechanisms under the IGA, and under the memoranda of understanding between NASA and ESA, NASA and the Government of Japan (GOJ), and NASA and MOSST. The IGA, and these memoranda are attached for reference but are not part of this Agreement. Neither the Russian Federation nor RSA is a party to the IGA, or these memoranda.

1.4 The Parties intend to proceed expeditiously to define their respective contributions to the Space Station Program, as well as operation and utilization concepts, preparatory to concluding a NASA-RSA Memorandum of Understanding (MOU) covering their entire cooperation in the program. In addition, the Parties note that the Government of the Russian Federation and the Partners stated on March 18, 1994, their intention to negotiate the following agreements: a Protocol to amend the IGA so that Russia may become a party to it and to provide for Russian participation in the Space Station Program as a partner; and a provisional arrangement concerning application of the IGA and Protocol pending entry into force of the Protocol. The Parties envision that this Agreement will remain in effect until the NASA-RSA MOU and Protocol have entered into force.

■ Article 2—Responsibilities

2.1   While undertaking activities under this Agreement, NASA will provide overall program coordination and direction and perform overall system engineering and integration for the Space Station. Boeing Aerospace is the U.S. prime contractor for system engineering and integration, and as such will assist NASA as required in these activities. RSA will provide overall development, coordination, management, and systems engineering and integration for its Space Station elements. RSA will participate in the management of the program and in the overall Space Station system engineering and integration. NPO Energia is the Russian prime contractor for system engineering and integration, and as such will assist RSA as required in these activities. NASA and RSA each remains ultimately responsible for performance of responsibilities delegated to its respective prime contractor.

2.2   NASA will conduct technical and managerial reviews of the Space Station program with RSA and the Cooperating Agencies of the Space Station Partners, as appropriate. NASA and RSA will develop all necessary joint documentation required for efficient execution of activities under this Agreement. RSA will participate with NASA and the Cooperating Agencies of the Space Station Partners in the management bodies as provided in Article 3.

■ Article 3—Management

3.1   RSA is responsible for management of its activities in accordance with this Agreement, and NASA is responsible for management of its activities in accordance with this Agreement, the IGA and the memoranda of understanding between NASA and the Cooperating Agencies of the Space Station Partners, and implementing arrangements under the IGA and the memoranda of understanding. Program management activities during the initial cooperation under this Agreement will be consistent with the Addendum.

3.2   The NASA Space Station Program Director at NASA Headquarters and the RSA Deputy General Director in Moscow will be responsible for their respective activities.

3.3   The NASA Space Station Program Manager at the Johnson Space Center and the RSA Deputy Division Chief in Moscow will implement their respective activities under the direction of their respective Agencies.

3.4   For initial cooperation under this Agreement, in accordance with Article 1, this Article establishes the management mechanisms to coordinate the respective design and development activities of NASA and RSA, to establish applicable requirements, to assure appropriate technical, operational, utilization, safety, and other activities, to establish interfaces between the Space Station elements, to review decisions, to establish schedules, to review the status of the activities, to report progress and to resolve issues and disputes as they may arise.

3.5   The NASA-RSA Program Coordination Committee (PCC), co-chaired by the NASA Space Station Program Director and the RSA Deputy General Director, will meet periodically or at the request of either Party to review the Parties' respective activities. The Co-Chairmen will together take those decisions necessary to assure implementation of the cooperative activities related to Space Station flight elements and to Space Station-unique ground elements provided by the Parties. In taking decisions regarding design and development, the NASA-RSA PCC will consider operation and utilization impacts, and will also consider design and development recommendations from the Multilateral Coordination Board (MCB) described below. However, decisions regarding operation and utilization activities will be taken by the MCB. The NASA-RSA PCC Co-Chairmen will decide on the location and timing of the meetings. If the Co-Chairmen agree that a specific issue or decision requires consideration by a Cooperating Agency of the Space Station Partners at the PCC level, the NASA-RSA PCC may meet jointly with the NASA-ESA PCC, the NASA-GOJ PCC, and/or the NASA-CSA PCC.

3.6   Space Station requirements, configuration, housekeeping resource allocations for design purposes, and element interfaces; Space Station activities through the completion of assembly and initial operational verification and other Space Station configuration control activities will be controlled by the Space Station Control Board (SSCB) chaired by NASA. RSA will participate on an equal basis with members of the SSCB and on such subordinate boards thereof as may be agreed, attending and participating when these boards consider items which affect the RSA-provided elements, interfaces between the NASA-provided and the RSA-provided elements, and interfaces between the RSA-provided elements and elements provided by the Cooperating Agencies of Space Station Partners. Decisions by the SSCB Chairman may be appealed to the NASA-RSA PCC, although it is the duty of the SSCB Chairman to make every effort to reach consensus with RSA rather than have RSA refer issues to the NASA-RSA PCC. Such appeals will be made and processed expeditiously. Pending resolution of appeals, RSA need not proceed with the implementation of a SSCB decision as far as its provided elements are concerned; NASA may, however, proceed with a SSCB decision as far as its provided elements are concerned. NASA will participate on RSA Space Station control boards, and on such subordinate boards thereof as may be agreed, attending as appropriate.

3.7   The Space Station System Specification and any modifications thereto, signed by the NASA Space Station Program Manager, the RSA Deputy General Director, and their counterparts in the Cooperating Agencies of the Space Station Partners, and approved by the SSCB, contain the requirements related to elements provided by the Parties, and the Cooperating Agencies of the Space Station Partners.

3.8   The Parties will work through the above management mechanisms to seek agreement on a case-by-case basis with the intention to use interchangeable hardware and software to the maximum extent possible in order to promote efficient and effective Space Station operations, including reducing the burden on the Space Station logistics system.

3.9   The NASA Space Station Program Office and the RSA Division for Manned Space Flight are responsible for NASA-RSA technical liaison activities. In order to facilitate the working relationships between the NASA Program Office in Houston and RSA, RSA will provide, and NASA will accommodate, the RSA liaison to the NASA Space Station Program Office. Similarly, NASA will provide and RSA will provide support for accommodation of the NASA liaison to the RSA in Moscow. RSA may also provide additional representative(s) to NASA Headquarters in Washington, DC, to further facilitate the program working relationships. Arrangements specifying conditions relating to the liaison relationships will be agreed to by the Co-Chairmen of the NASA-RSA PCC.

3.10 RSA will participate in selected NASA reviews on Space Station requirements, architecture and interfaces. Similarly, NASA will participate in selected RSA reviews; the Cooperating Agencies of the Space Station Partners will participate as appropriate.

3.11 Party has responsibilities regarding the management of its respective operations and utilization activities and the overall Space Station operations and utilization activities, in accordance with the provisions of this Agreement. Activities under this agreement will comprise long-range planning and top-level direction and coordination which will be performed by the strategic-level organizations, and which will be consistent with the Addendum. Operations plans will be developed by the Parties. These plans will include any necessary contingency plans for the safe and efficient operation of the Space Station while on-orbit. They will also outline the division of responsibilities of the Parties, taking into account RSA's particular operations capabilities during Phase 2, in the framework of a unified command and control center concept as outlined in the Addendum.

3.12 The Multilateral Coordination Board (MCB), an established Space Station management body, meets periodically or promptly at the request of the Parties or a Cooperating Agency of the Space Station Partners with the task to ensure coordination of activities related to the operation and utilization of the Space Station. The Parties to this Agreement and the Cooperating Agencies of the Space Station Partners will plan and coordinate activities affecting the safe, efficient and effective operation and utilization of the Space Station through the MCB, except as otherwise specifically provided in this Agreement. The MCB will comprise the NASA Space Station Program Director; the RSA Deputy General Director; the STA Director-General of the Research and Development Bureau; the ESA Columbus Programme Department Head; and the CSA Vice President for Human Spaceflight. The NASA Space Station Program Director will chair the MCB. The Parties agree that all MCB decisions should be made by consensus. However, where consensus cannot be achieved on any specific issue within the purview of the MCB within the time required, the issue will be resolved on the basis of the principles which govern the MCB.

3.13 The MCB has established Panels responsible for the long-range strategic coordination of the operation and utilization of the Space Station, called the System Operations Panel (SOP) and the User Operations Panel (UOP) respectively. The MCB approves, on an annual basis, a Consolidated Operations and Utilization Plan (COUP) for the Space Station based on the annual Composite Operations Plan and the annual Composite Utilization Plan developed by the Panels. In doing so, the MCB will be responsible for resolving any conflicts between the Composite Operations Plan and the Composite Utilization Plan which cannot be resolved by the Panels. The COUP will be prepared by the User Operations Panel and agreed to by the System Operations Panel. The COUP will be implemented by the appropriate tactical and execution-level organizations. Any portions of a COUP which cover activities prior to Assembly Complete plus one year of initial operational verification will be subject to adjustments by the SSCB that are required to assemble, verify, operate and maintain the Space Station.

3.14 The Parties will use their best efforts, in consultation with the Cooperating Agencies of the Space Station Partners, to incorporate any necessary changes in management operation within the framework of the management structure to reflect the expanded number of partners in the Space Station program.

▌ Article 4—Safety and Mission Assurance

4.1 In order to assure safety, NASA has the responsibility, working with the RSA and the Cooperating Agencies of the Space Station Partners, to establish overall Space Station safety and mission assurance requirements and plans.

4.2 RSA will develop detailed safety and mission assurance requirements and plans, using its own requirements for its Space Station hardware and software. Such requirements and plans must meet or exceed the overall Space Station safety requirements and plans. Requirements for which meet or exceed criteria are not appropriate will be determined by agreement of the Parties. RSA will have the responsibility to implement Space Station safety and mission assurance requirements and plans with respect to the elements and payloads it provides throughout the lifetime of the program, and to certify that such requirements and plans have been met. NASA will have the overall responsibility to certify that all Space Station elements and payloads are safe.

4.3 The Parties will exchange information necessary in order to conduct system safety reviews. The Parties will also conduct safety reviews of the elements and payloads they provide.

■ Article 5—Cross-Waiver of Liability

5.1 The objective of this Article is to establish a cross-waiver of liability by the Parties to this Agreement and related entities in the interest of encouraging participation in space exploration, use and investment through the Space Station. In addition, in light of the liability requirements in Article 16 of the IGA, a second purpose of this article is to fulfill the obligation of the United States of America, as a Partner State, to extend the cross-waiver to related entities of the United States Government in the Space Station Program. Thus, pursuant to this Article, RSA, as a related entity of NASA and the Government of the United States of America, for purposes of this article, is protected by application of the Cross-Waiver of Liability agreed to by the Partner States in the IGA. The cross-waiver of liability shall be broadly construed to achieve the objective of encouraging participation in space activities.

5.2 For the Purposes of this Article:

(a) The term "damage" means:

(1) bodily injury to, or other impairment of health of, or death of, any person;

(2) damage to, loss of, or loss of use of any property;

(3) loss of revenue or profits; or

(4) other direct, indirect, or consequential damage.

(b) The term "launch vehicle" means an object (or any part thereof) intended for launch, launched from Earth, or returning to Earth which carries payloads or persons, or both.

(c) "Partner State" means a signatory to the IGA. A "Partner State" includes its Cooperating Agency. It also includes any entity specified in the Memorandum of Understanding between NASA and the Government of Japan to assist the Government of Japan's Cooperating Agency in the implementation of that Agreement.

(d) The term "payload" means all property to be flown or used on or in a launch vehicle or the Space Station.

(e) The term "Protected Space Operations" means all launch vehicle activities, Space Station activities, and payload activities on Earth, in outer space, or in transit between Earth and outer space done in implementation of this Agreement, the IGA, the memoranda of understanding between NASA and the Cooperating Agencies of the Space Station Partners, or implementing arrangements under the IGA and the memoranda of understanding. It includes, but is not limited to:

(1) Research, design, development, test, manufacture, assembly, integration, operation, or use of launch or transfer vehicles (for example, the orbital maneuvering vehicle), the Space Station, or a payload, as well as related support equipment and facilities and services;

(2) All activities related to ground support, test, training, simulation, or guidance and control equipment, and related facilities or services.

"Protected Space Operations" also includes all activities related to evolution of the Space Station, as provided for in Article 14 of the IGA. "Protected Space Operations" excludes activities on Earth which are conducted on return from the Space Station to develop further a payload's product or process for use other than for Space Station-related activities in implementation of this Agreement or the IGA.

(f) The term "related entity" means:

(1) A contractor or subcontractor of a Party or a Partner State at any tier;

(2) A user or customer of a Party or a Partner State at any tier; or

(3) A contractor or subcontractor of a user or customer of a Party or a Partner State at any tier.

"Contractors" and "subcontractors" include suppliers of any kind.

5.3 (a) Each Party agrees to a cross-waiver of liability pursuant to which each Party waives all claims against any of the entities or persons listed in paragraphs 5.3(a)(1) through 5.3(a)(4) below based on damage arising out of Protected Space Operations. This cross-waiver shall apply only if the person, entity, or property causing the damage is involved in Protected Space Operations and the person, entity, or property damaged is damaged by virtue of its involvement in Protected Space Operations. The cross-waiver shall apply to any claims for damage, whatever the legal basis for such claims, including but not limited to delict and tort (including negligence of every degree and kind) and contract, against:

(1) the other Party;

(2) a Partner State other than the United States of America;

(3) a related entity of any entity identified in subparagraphs 5.3(1)(1) or 5.3(a)(2) above; or

(4) the employees of any of the entities identified in subparagraphs 5.3(a)(1) through 5.3(a)(3) above.

(b) In addition, each Party shall extend the cross-waiver of liability as set forth in paragraph 5.3(a) above to its own related entities by requiring them, by contract or otherwise, to agree to waive all claims against the entities or persons identified in subparagraphs 5.3(a)(1) through 5.3(a)(4) above.

(c) For avoidance of doubt, this cross-waiver of liability includes a cross-waiver of liability arising from the Convention on International Liability for Damage Caused by Space Objects, of March 29, 1972, where the person, entity, or property causing the damage is involved in Protected Space Operations and the person, entity, or property damaged is damaged by virtue of its involvement in Protected Space Operations.

(d) Notwithstanding the other provisions of this Article, this cross-waiver of liability shall not be applicable to:

(1) claims between NASA and RSA arising out of activities conducted under any contract between NASA and RSA;

(2) claims between a Party and its other related entities or between its own related entities;

(3) claims made by a natural person, his/her estate, survivors, or subrogees for injury or death of such natural person;

(4) claims for damage caused by willful misconduct;

(5) intellectual property claims.

(e) Nothing in this Article shall be construed to create the basis for a claim or suit where none would otherwise exist.

▌ Article 6—Exchange of Technical Data and Goods

6.1 Except as otherwise provided in this Article, each Party will transfer all technical data and goods considered to be necessary (by both parties to any transfer) to fulfill its respective responsibilities under this Agreement. In addition, NASA may request a Cooperating Agency of a Space Station Partner to transfer directly to RSA technical data and goods necessary to fulfill NASA's responsibilities under this Agreement. NASA may also request RSA to transfer directly to a Cooperating Agency of a Space Station Partner technical data and goods necessary to fulfill RSA's responsibilities under this Agreement. Each Party undertakes to handle expeditiously any request for technical data or goods presented by the other Party for the purposes of this cooperation. This paragraph will not require either Party to transfer any technical data and goods in contravention of its national laws or regulations.

6.2 The transfers of technical data and goods under this Agreement will be subject to the restrictions set forth in this paragraph. Technical data and goods not covered by these restrictions will be transferred without restrictions, except as otherwise restricted by national laws and regulations.

    (a) The furnishing Party or a Cooperating Agency of a Space Station Partner will mark with a notice, or otherwise specifically identify, the technical data or goods that are to be protected for export control purposes. Such notice or identification will indicate any specific conditions regarding how such technical data or goods may be used by the receiving Party and its contractors and subcontractors, and by the Cooperating Agency of a Space Station Partner and its contractors and subcontractors. These conditions will include: (1) that such technical data or goods will be used only for the international Space Station program to fulfill responsibilities of the Parties or of a Cooperating Agency of a Space Station Partner, and (2) that such technical data or goods will not be used by persons or entities other than the receiving Party, its contractors or subcontractors, and by the Cooperating Agency of a Space Station Partner, its contractors or subcontractors, or for any other purpose without the prior written permission of the furnishing Party.

    (b) The furnishing Party or a Cooperating Agency of a Space Station Partner will mark with a notice the technical data that are to be protected for proprietary rights purposes. Such notice will indicate any specific conditions regarding how such technical data may be used by the receiving Party and its contractors and subcontractors, and by the Cooperating Agency of a Space Station Partner and its contractors and subcontractors, including (1) that such technical data will be used, duplicated, or disclosed only for the international Space Station program to fulfill responsibilities of the Parties or of a Cooperating Agency of the Space Station Partner, and (2) that such technical data will not be used by persons or entities other than the receiving Party, its contractors or subcontractors, the Cooperating Agency of a Space Station Partner, its contractors or subcontractors, or for any other purpose without the prior written permission of the furnishing Party.

6.3 Each Party will take all necessary steps to ensure that technical data and goods received by it under subparagraph 6.2(a) or 6.2(b) above will be treated by the Receiving Party, and other persons and entities (including Cooperating Agencies of the Space Station Partners, contractors and subcontractors) to which the data and goods are subsequently transferred in accordance with the terms and conditions of the notice. (The Cooperating Agencies of the Space Station Partners and their respective Governments have obligations under the IGA to protect data and goods transferred by RSA under this Agreement.) Each Party will take all reasonably necessary steps, including ensuring appropriate contractual conditions in their contracts and subcontracts, to prevent unauthorized use, disclosure, or retransfer of, or unauthorized access to, such technical data and goods.

6.4   It is not the intent of the Parties to grant, through this Agreement, any rights to a recipient beyond the right to use, disclose, or retransfer received technical data or goods consistent with conditions imposed under this Article.

6.5   For purposes of this cooperation, interface, integration, safety and testing data (excluding detailed design, manufacturing and processing data, and associated software) shall be exchanged by the Parties without restrictions as to use or disclosure, except as specifically required by national laws and regulations relating to export controls.

■ Article 7—Intellectual Property

7.1   With the exception of the intellectual property rights referred to in Article 6, Exchange of Technical Data and Goods, and subject to national laws and regulations, provisions for the protection and allocation of intellectual property rights created during the course of cooperation between the Parties to this Agreement are set forth in Annex 1 of the June 17, 1992, Agreement between the United States of America and the Russian Federation Concerning Cooperation in the Exploration and Use of Outer Space for Peaceful Purposes.

7.2   Except as set forth in paragraph 7.1, nothing in this Agreement will be construed as granting or implying any rights to, or interest in, patents or inventions of the Parties or their contractors or subcontractors.

■ Article 8—Public Information

Each Party will coordinate, as appropriate, with the other in advance concerning its own or joint public information activities related to subjects covered by this Agreement.

■ Article 9—Customs and Immigration

9.1   Each Party will use its best efforts to facilitate the movement of persons and goods necessary to implement this Agreement, into and out of its territory, subject to laws and regulations of its respective country.

9.2   Subject to its respective countries' laws and regulations, each Party will use its best efforts to facilitate provision of the appropriate entry and residence documentation for the other Party's nationals and their families, or for the nationals and their families of Space Station Partner States who enter, exit or reside within its territory in order to carry out activities described herein.

9.3   The Parties will use their best efforts to arrange in their respective countries for free customs clearance, to include no payment of import and export duties and no payment for the conduct of customs procedures, for entrance to, and exit from, their respective countries, for goods required for implementation of the activities described herein.

9.4   RSA will take steps to facilitate the movement of persons and goods and clearances to and from launch facilities RSA will utilize to fulfill its obligations under this Agreement.

■ Article 10—Financial Arrangements

10.1   Each Party will bear the costs of fulfilling its responsibilities, including but not limited to costs of compensation, travel and subsistence of its own personnel and transportation of all equipment and other items for which it is responsible under this Agreement, except as provided for in contractual or other arrangements between the Parties.

10.2 The financial obligations of each Party pursuant to this Agreement are subject to its funding procedures and the availability of appropriated funds. Recognizing the importance of Space Station cooperation, each Party undertakes to make its best efforts to obtain approval for the funds to meet those obligations, consistent with its respective funding procedures.

10.3 In the event that funding problems arise that may affect a Party's ability to fulfill its responsibilities under this Agreement, that Party will promptly notify and consult with the other Party.

10.4 The Parties will seek to minimize the exchange of funds while carrying out their respective responsibilities in this cooperative program, including, if they agree, through the use of barter, that is, the provision of goods or services.

■ Article 11—Termination

11.1 This Agreement may be terminated at any time by giving at least three months prior notice by diplomatic note. Upon notice of termination for any reason, NASA and RSA will expeditiously negotiate an agreement concerning the terms and conditions of termination. To the extent that termination affects specific rights or obligations of a Cooperating Agency of a Space Station Partner under the IGA or the MOU between NASA and that Cooperating Agency, NASA will consult with the affected Cooperating Agency before concluding any such agreement.

11.2 Termination by either Party will not affect that Party's continuing rights and obligations under this Agreement with regard to liability and the protection of technical data and goods unless otherwise agreed in a termination agreement pursuant to Article 11.1.

■ Article 12—Amendment

This Agreement may be amended by written agreement of the Parties.

■ Article 13—Language

The working language for activities under this Agreement will be the English language and data and information generated or provided under this Agreement will be in the English language.

■ Article 14—Entry into Force

This Agreement will enter into force upon the exchange of diplomatic notes confirming its terms by the Government of the United States of America and the Government of the Russian Federation. Unless this Agreement is terminated pursuant to Article 11, it will remain in effect consistent with Article 1.4, until otherwise agreed by the Parties.

Done at Washington, in duplicate, this twenty-third day of June, 1994, in the English and Russian languages, each text being equally authentic.

FOR THE NATIONAL AERONAUTICS AND
SPACE ADMINISTRATION OF THE
UNITED STATES OF AMERICA: FOR THE RUSSIAN SPACE AGENCY:

Dan Goldin   Yuri Koptev

Appendix B:
Public Sector
U. S.-Russian
Cooperative
Projects | B

Public Sector U.S.-Russian Cooperative Projects

Field and mission/title	Description	Russian entity	Status
Astrophysics			
Exchange of Compton and Granat Data	Mission timelines being exchanged. U.S. proposal to perform correlative studies of neutron stars and stellar-mass black hole candidates using BATSE, SIGMA, and ART-P.	Institute of Space Research (IKI)	Granat expected to operate in three-axis stabilized mode through 10/94. Russian and French partners considering operating in a scanning mode through about 10/95. Scientific data exchange discussed at 4/94 JWG.
Experiment for gamma-ray and optical transients (EGOTS)/ Konus-A	Russian proposal for U.S. participation in proposed flight in 1995.	Ioffe Institute, St. Petersburg	Approved by RSA as Konus-A. To be flown on a low-Earth orbit spacecraft scheduled for launch in early 1995. Could be reflown in 1-2 years as Konus-A2.
Gamma-ray burst data exchange	Exchange of data among BATSE, Granat, and Ulysses.	IKI	Burst timing/location now or will be available from PVO, SMM, Ginga, Phobos-2, Compton, WIND, Ulysses, Granat, Coronas-1 and Mars '94 (now '96). Good exchange of data currently among BATSE, Granat and Ulysses.
HETE	International gamma-ray-burst survey using a small satellite to be built by the United States, in collaboration with France and Japan; NASA to loan ground-station equipment to Russia to receive signals from HETE and to alert visible-light observatories to look at burst sources.	Pulkovo Observatory, St. Petersburg	Satellite launch planned for 7/95 at the earliest. Loan agreement to be written.
Long-duration balloon studies for Relict-3 mission	Russian Relict-3 mission funded for Phase A. NASA participation under consideration.	IKI	Sides agreed that long-duration balloon-based observations might be a good preliminary step. Russians noted that LDBF near Moscow had been closed due to funding problems. NASA to discuss with its science community the usefulness of such flights in northern hemisphere.
Operations and archiving for Spectrum-X-Gamma (SXG)	Cooperation in archiving and operations for Spectrum-X, using NASA-supplied hardware for the SXG archiving system.	IKI	Smithsonian Astrophysical Observatory selected as the U.S. SXG Coordinating Facility, to support U.S. guest observers and work with IKI to define and procure archiving hardware.

Precision Gamma-ray spectrometer (PGS)	U.S. germanium gamma-ray detectors to fly as part of Russian experiment on Mars '94 (now '96) orbiter; U.S. contribution sponsored by NASA and DOE; being handled in Solar System Exploration JWG.	IKI	Launch date slipped to 1996. U.S. components, built by DOE/LANL, shipped to IKI in 1993.
Radioastron	Russian mission; NASA to provide DSN support, loan of VLBA recording terminals, and VLBA observing time and to cooperate in correlation of selected data sets.	IKI, Russian Academy of Sciences (RAS)	Launch date TBD, no earlier than 1997 (following Spectrum-X-Gamma); engineering model of 10-m deployable antenna being fabricated, scheduled to be completed by the end of 1994; ground test early 1995 at Puschino Radio Observatory.
Relict-2	Russian mission similar to COBE. NASA participation in development and fabrication of receivers.	IKI	Development of flight model spacecraft, utilizing Prognosz engineering model, pending availability of funds; launch scheduled for 12/95. Computer workstation and software for processing COBE data by Relict science team procured at GSFC; to be loaned to IKI by NASA.
Spectrum-UV	170-cm telescope for imaging and spectroscopy, with possible co-aligned smaller telescopes, in a 7-day, highly elliptical orbit.	Institute of Astronomy, RAS, IKI	Currently not scheduled for flight. Ukraine, Canada, Germany and Italy also reportedly involved.
Spectrum-X-Gamma	NASA to provide Stellar X-ray Polarimeter (SXRP), Monitoring X-ray Experiment (MOXE—All Sky X-ray Monitor), and filters for Russian EUVITA instrument.	IKI	Scheduled launch date is late 1996. Engineering models of SXRP and MOXE instruments accepted by IKI; flight units in production. MOU being developed.
Wind/Konus	Gamma-ray burst detectors for Wind spacecraft, as part of ISTP; co-investigator from the Ioffe Institute.	Ioffe Institute	Instrument delivered and integrated on spacecraft. Launched 11/1/94.
Earth Science and Environmental Monitoring JWGs			
BOREAS Field Experiment	Joint U.S.-Canada experiment on interactions between the boreal forest and the atmosphere.	Institute of Forest and Timber (IFT), Siberian Branch, RAS	Russian investigator selected through peer-reviewed process, resident at the University of New Hampshire.
Correlative measurements of ozone	Cooperation involving ground-, balloon-, and aircraft-based measurements correlative with space-based measurements.	Institute of Atmospheric Optics (IAO)	Joint Implementation Team formed 4/94; more than 30 Russian proposals currently under study.

Public Sector U.S.-Russian Cooperative Projects (Cont'd)

Field and mission/title	Description	Russian entity	Status
Crustal Deformation in Pamir–Tien Shan	Determining the mechanisms of mountain building using the Tien Shan mountains as a natural laboratory.	Institute of High Temperature Physics, RAS	Major field program during summers 1993 and 1994; included scientists from Russia, Kyrgyzstan, Kazakhstan, and the United States; further field experiments planned for summer 1995.
Earthquake Precursors Study	Data exchanges in the area of atmospheric precursors to earthquakes.	Institute of Astronomy, RAS	Some exchanges completed; further exchanges, including retrospective analysis of satellite data for selected California earthquakes, to continue.
FEDMAC–Sayani Field Experiments	Study of forest health using in situ and satellite data.	IFT	Field work complete; final publications being prepared.
FIFE–Kursk Field Experiments	Study of climatologically significant land–surface parameters using satellite data.	Institute of Computational Mathematics, RAS	Field work complete; final publications being prepared.
Gravity and Magnetics in Tibet and China	Analysis of gravity data from Russia and western China by bilateral investigator groups.	Institute of Mathematical Geophysics, RAS	Modeling continuing.
Internet Connectivity	Extending electronic communications via the existing NASA Science Internet (NSI) connection with IKI to reach Russian Earth science facilities.	IKI, RAS	List of priority sites to be jointly developed and provided to NSI for implementation.
Kamchatka Volcanological Studies	Exchange of U.S. and Russian aircraft data and joint ground measurements; joint analysis of data.	Institute of Geology, Petrology, Mineralogy, and Geochemistry (IGPMG), RAS	Learjet overflights completed 8-9/94; additional aircraft flights under consideration; data exchanges under way.
LITE Shuttle mission	To coordinate Russian ground-based LIDAR measurements with Shuttle-based measurements; Russian principal investigator on science team.	IAO, Siberian Branch, RAS	9/94 Shuttle flight.
Meteor-3/TOMS	Flight of NASA Total Ozone Mapping Spectrometer (TOMS) instrument on Meteor-3 polar orbiter.	Russian Federal Service for Hydrometeorology and Environmental Monitoring (ROSHYDROMET)	8/91 launch; TOMS instrument failed in early 1995.

Project	Description	Institution	Status
Moz-Obzor (Priroda)	Use of the Moz-Obzor ocean-color instrument on Priroda in conjunction with SeaWiFS.	Institute of Radioengineering and Electronics (IRE)	U.S. proposal accepted by the Priroda Science Team.
Operational Data Transmissions	Processing of limited amount of NOAA satellite products by ROSHYDROMET for use in local and regional forecasting; provided by NOAA through GTS.	ROSHYDROMET	Under way
Priroda	NASA invited to participate in the science associated with the Mir-Priroda module.	IRE	Seven NASA investigations using Russian-provided instruments; launch December 1995.
Regional Tectonics in Eurasia	Through WEGENER program.	Institute of Geology, RAS	GPS survey of Caucasus completed 1993; workshop 6/94 in St. Petersburg.
SAGE Flight	Flight of SAGE instrument on a Russian Meteor-3M as part of NASA EOS Program.	RSA, Scientific Research Institute of Electromechanics (NIIEM)	Agreed at 12/94 GCC meeting.
SeaWiFS	Variety of activities involving Russian participation in U.S. SeaWiFS project.	Shirshov Institute, RAS	Russian scientist selected as a principal investigator in the SeaWiFS program; Russian bio-optical data delivered to SeaWiFS database; NASA to facilitate further Russian scientist participation.
Siberian AVHRR Stations	Installation of two NASA-provided HRPT stations at Yakutsk and Khabarovsk to support IGBP 1-km data set project.	ROSHYDROMET	Installed 1-2/95.
Space Geodetic Measurements	Long-term loan of Mark 3 VLBI data-acquisition systems; satellite laser tracking and exchanges of data; potential project in atmospheric precursors for earthquakes.	Institute of Astronomy, RAS	Loan agreement in place; first VLBI experiments summer 1994; satellite laser ranging (SLR) data exchanges well-established.
TAIGA Study of the Boreal Forest	Exchange of Russian ground truth and data from NASA-provided satellite HRPT receiving station in Krasnoyarsk, to study forest productivity, forest health, fire risk, and fire history in the context of the global carbon cycle.	International Forestry Institute (IFI), RAS	Letter agreement concluded 3/94; equipment installed 11/94.
TOMS Flight	Flight of another TOMS on a Russian Meteor-3M.	RSA, NIIEM	Agreed at 12/94 GCC meeting.
Watershed Hydrology Project (Priroda)	Study of microwave remote sensing for large-watershed hydrology.	IRE, RAS	Formalized at 6/94 Priroda Scientific Council meeting.

Public Sector U.S.-Russian Cooperative Projects (Cont'd)

Field and mission/title	Description	Russian entity	Status
Solar System Exploration			
Antarctic instrument cross-calibration balloon flight	To cross-calibrate remote-sensing gamma-ray and neutron spectrometers for geochemical observation of planetary surfaces.	IKI	At the 10/94 JWG, discussed successful Antarctic balloon flight; recommended expansion of goals to encompass intercalibration of geochemical systems for U.S. and Russian planetary missions; detailed plan to be submitted at next JWG meeting.
Coordination and exchange of data for science exploration of Venus	Data exchange from the Magellan, Galileo, and Venera/Vega missions; data documentation and archiving; review of Russian scientists' experience in the Magellan Guest Investigator Program; specific joint investigations and studies; documentation of future exploration goals, including consideration of concepts for future techniques and experiments to achieve these goals; selection of future landing sites and/or balloon traverses; and identification of future joint/complementary experiments and missions.	IKI	Discussed at 10/94 JWG; progress satisfactory; meetings 1/95 and 3/95 in Arizona and Houston, TX.
Coordination of future missions	Technical study of future cooperative solar system exploration missions.	IKI	10/94 JWG meeting received reports from joint technical study teams for Mars Together, and Fire and Ice. JWG endorsed reports in principle and forwarded them to the 12/94, GCC meeting, at which the principals asked the JWG to continue the study activity and produce a specific recommendation at the next GCC meeting (6/95).
Coordination of missions	U.S. VLBI tracking of Mars '96 during cruise to Mars; U.S. Mars Surveyor relay of Mars '96 lander data to verify lander operability; joint tracking campaigns during Mars orbit phase; and joint U.S.-Russian VLBI tracking experiments.	IKI, RAS	Discussed at 10/94 JWG meeting; progress satisfactory.

Coordination of science observations and exchanges of data	IKI	Development of common data formats; systematic exchange of data sets in the agreed formats; and exchange of participating scientists.	Activities involving the Phobos, Mars '96-'98, Mars Surveyor, and Pathfinder missions reviewed at 10/94 JWG meeting; substantial progress made. Detailed status of each Participating Scientist to be reviewed at the next JWG meeting.
Exobiology	IKI	Characterization of Mars sites of interest to exobiology, Mars mission strategies, instrumentation for various missions, and planetary protection.	At the 10/94 JWG meeting, implementation team recommended continuation of joint Mars site studies and a joint workshop on planetary protection measures for Mars sample return missions.
Ground-based observations in support of planetary missions	IKI	To coordinate Mars Watch and Near-Earth Objects Watch.	At the 10/94 JWG, agreed to continue and strengthen joint ground-based observing programs through an exchange of observing plans and results.
Mars '96	IKI, RAS	Fly two copies of U.S. Mars Oxidant (MOX) experiment on Mars '96 landers.	Flight postponed to 1996; progress satisfactory.
Mars engineering models	IKI	To develop realistic models of the Martian near-surface wind environment to support future lander missions.	United States to continue modeling work; both sides to seek better understanding and verification of the results by comparison with available observations and other models.
Mars landing-site selection	IKI	To develop models of the Martian surface for the design of future missions and selection of landing sites.	Agreement at the 10/94 JWG meeting on several specific steps to develop additional information for refinement of the engineering model; met in 3/95 in Houston to review potential landing sites for Mars '96 small landers and penetrators.
Mercuric iodide room temperature x-ray detecting system	Max Planck Institute, Germany; IKI	Part of German alpha backscatter instrument for Mars '96.	Hardware delivered; awaiting 1996 launch.
Space Biomedicine, Life Support Systems and Microgravity Sciences			
Biological investigations aboard Mir	Institute of Biomedical Problems (IBMP)	Investigations to include "Seed to Seed" experiment with dwarf wheat and investigations of avian egg development.	At 3-4/94 JWG meeting, two sides agreed to proceed with implementation.

Public Sector U.S.-Russian Cooperative Projects (Cont'd)

Field and mission/title	Description	Russian entity	Status
Bion 11 and 12	U.S. reimbursable participation in primate scientific programs on the Russian Bion 11 and 12 biosatellite missions, scheduled for 1996 and 1998.	RSA, IBMP, Central Specialized Design Bureau, Samara (TsKB)	Contract signed 12/94; series of working meetings under way at NASA Ames Research Center (ARC) and IBMP.
Bion 10 flight experiments	Flight of U.S. primate experiments on Russian biosatellite mission.	IBMP, Institute of Evolutionary Physiology and Biochemistry (IEPB), RAS	Experiments completed; analysis of results presented and published.
Cooperation regarding space-radiation-environment databases	Beginning with 9/94 meeting, pursue a systematic exchange of databases.	IEPB	At 9/94 meeting, two sides agreed to begin identifying sources of data from the respective countries, types of data, and database formats. Specific range of data types sought agreed to at 3-4/94 JWG meeting.
Flight of TEPC (Tissue Equivalent Proportional Counter) on the Mars '94 mission	U.S.-provided TEPC to be flown on the Mars '94 mission (now slipped to 1996).	IEPB	Agreement in principle; implementing agreement under negotiation.
IBMP life-support testbed upgrade	Russia invited the United States to take part in upgrading the IBMP life-support testbed in order to set up an international center on the development and testing of complex physical-chemical and environmental life-support systems.	IBMP	At 3-4/94 JWG meeting, U.S. side agreed to study the proposal.
Joint experimental dosimetric measurements		IEPB	Measurements on STS-60 mission completed; to continue on subsequent flights.
Joint rodent developmental experiment	Joint experiment on board the Space Shuttle in 10/94.	IBMP	Agreed during 3-4/94 JWG meeting; preparations under way.
Medical systems in support of the International Space Station	Testing and evaluation of medical programs during Phase One.	IBMP, Cosmonaut Training Center (TsPK), RSA	Ongoing.
Publication of "Foundations of Space Biology and Medicine"	Multivolume compendium of U.S. and Russian articles on research in the field.	IBMP, Ministry of Health	Exchange of chapters for Volume III completed; for Volume IV, to be completed by 9/30/95.

Radiation-exposure standards	Exchange of information on current U.S. and Russian standards and exploration of possibility of convergence on one process and one set of standards.	IEPB	Forum to be established; first meeting in 9/94.
Shuttle/Mir program	Fundamental and applied medical and physiological experiments aboard Mir and Shuttle.	IBMP, others	Being implemented.
SLS-2 flight experiments	Joint U.S.-Russian experiments in biology.	IBMP	Final report released 9/94. Russian specialists invited to a symposium in the United States on SLS-2, held fall 1994.
Space-crew safety, operational efficiency and unified U.S.-Russian medical support in piloted missions	Updating and advancing medical issues of space crew safety and operational efficiency; establishing a unified U.S.-Russian system for medical support in piloted missions.	IBMP, TsPK, Ministry of Defense, RSA	Ongoing.
Standardizing techniques for physical, chemical, and biological analyses of recovered water and air	Coordination of Russian and U.S. sampling operations and data sharing.	IBMP	At 2/95 meeting held between IBMP toxicologist and NASA, sampling methods agreed to for first Phase One flight.
Support of Russian scientific community under the NASA/RSA contract	Funding of Russian space scientists and technologists in eight discipline areas ($20 million set-aside).	RSA's Scientific and Technical Advisory Council, 50 scientific organizations	Joint meeting held in Moscow and the United States to review process.
Unified approach to environmental standards	Setting appropriate standards for Phase One program.	IBMP	At 2/95 meeting held between IBMP toxicologist and NASA, sampling methods agreed to for first Phase One flight.
Space Physics			
Anomalous cosmic rays	Three-point approved program involving investigation of trapped anomalous cosmic rays, coordinated measurements with SAMPEX, and investigation of the mean ionic charge state of solar energetic particles. German investigators also participating.	Scientific Research Institute of Nuclear Physics, Moscow State University (NIIJAF MGU)	Comparison of modeling and observations discussed during 4-5/94 JWG. Agreement on reduction and comparison of data from 7-8/93 Cosmos flight and SAMPEX to derive the mirror point distribution of trapped anomalous cosmic rays. United States to prepare joint publication on new data. Russia to provide additional data from three COSMOS flights during the last solar cycle.

Public Sector U.S.-Russian Cooperative Projects (Cont'd)

Field and mission/title	Description	Russian entity	Status
Coronas	U.S. scientists invited to participate in mission operations planning and subsequent data analysis of Coronas-I, as well as the planned Coronas-F and Foton missions, to study solar activity.	Institute for Earth Magnetism, Ionosphere and Radio Propagation of the Russian Academy of Sciences (IZMIRAN)	Coronas-I operational. Letter agreement finalized 1/23/95.
Flight dynamics	Update *IACG Handbook on Trajectories, Mission Design and Operations*; conduct mission design for Relict-2.	IKI	Updated handbook expected to ready for distribution 9/94; small joint team formed to develop an electronic version.
Geospace	Interball–Tail and Interball–Aurora spacecraft planned for launch in 10/94 and spring 1995, respectively.	IKI	In 4–5/94 JWG, agreement reached on U.S. scientist participation in data analyses. NASA to fund investigators (subject to Interball principal investigator agreement and Interball science team member involvement).
IACG-coordinated campaigns	Coordinated observing campaigns using existing spacecraft regarding flow of energy in the magnetotail, collisionless boundaries in space plasmas, and solar events and their manifestations in geospace.	IKI	First campaign, led by GSFC, in fall 1993. Second campaign to be led by ESA. Second and third campaigns not before 1996.
Long-duration balloons	To fly long-duration balloons carrying U.S. payloads between North America and Russia; NASA to provide balloons, launch, and tracking.	NIIJAF MGU	First flight planned for 6/95.
Magnetospheric modeling	Development of a mathematical framework for describing the different components of the Earth's distant magnetic field; assembling data from space, mainly observations of the magentic field, used for calibrating the mathematical representations.	IKI	Most of the work is being done by the Goddard group (Stern, Tsyganenko, et al.); significant new developments and publications have resulted. JWG (4–5/94) concluded that the implementation team had achieved its intended purpose and decided to terminate its activity.

Project	Activity	Participants	Status/Comments
Solar Probe	Provide science input to Elachi/Galeev technical team.	IKI, RAS, IZMIRAN	Technical team to prepare preliminary report by 8/94, final report by 11/94. Joint Science Steering Group formed at 4-5/94 JWG meeting. U.S. side hosted workshop on "Near-Sun Science" in summer 1994.
TREK	U.S. ultra-heavy cosmic-ray detectors flown on Mir; NASA supplied detectors and leading data-analysis efforts, while RSA providing launch, recovery, and collaboration in data analysis.	IKI	Three small stacks and one-third of the large stack have been returned to Earth from Mir. Plans are under way to retrieve the remaining two-thirds of the external collector in 1995 by a joint U.S./Russian EVA. Afanasiev is currently at University of California at Berkeley to participate in TREK data analysis for one year.

ARC = Ames Reseach Center
ART-P = Advanced Roentgen Telescope-Positioning
AVHRR =Advanced Very High Resolution Radiometer
BATSE = Burst and Transient Source Experiment
BOREAS = Boreal Ecosystem-Atmosphere Study
COBE = Cosmic Background Explorer
DOE = Department of Energy
DSN = Deep Space Network
EOS = Earth Observing System
EUVITA = Extreme Ultraviolet Imaging Telescope Array
EVA = Extra Vehicular Activity
FEDMAC = Forest Ecosystem Dynamics Multispectral Airborne Campaign
FIFE = First International Field Experiment

GCC = Gore-Chernomyrdin Commission
GPS = Global Positioning System
GSFC = Goddard Space Flight Center
GTS = Global Telecommunications System
HETE = High Energy Transient Experiment
HRPT = High Respolution Picture Transmission
IGBP = International Geosphere-Biosphere Program
ISTP = International Solar Terrestrial Physics
JWG = Joint Working Group
LANL = Los Alamos National Laboratory
LDBF = Long Duration Balloon Facility
LIDAR = Light Detection and Ranging
LITE = LIDAR In-space Technology Experiment
MOU = Memorandum of Understanding

NASA = National Aeronautics and Space Administration
NOAA = National Oceanographic and Atmospheric Administration
PVO = Pioneer Venus Orbiter
RSA = Russian Space Agency
SAMPEX = Solar, Anomalous, Magnetospheric Explorer
SeaWiFS = Sea-Viewing Wide Field Sensor
SIGMA = French gamma ray instrument
SMM = Solar Maximum Mission
VLBA = Very Long Baseline Array
VLBI = Very Long Baseline Interferometer
WEGENER = Working Group of European Geo-scientists for the Establishment of Networks for Earthquake Research

SOURCE: Office of Technology Assessment, 1995.

Appendix C: Private Sector U.S.-Russian Cooperative Projects | C

Private Sector U.S.-Russian Cooperative Projects

Title/description	Business entity and nature of relationship	Russian entity	U.S. entity	Status
LOX-augmented nuclear thermal rocket engine design	Joint design study.	Energopool (Consortium)	Aerojet; Babcock & Wilcox	Pending U.S. government policy decisions on nuclear propulsion.
NK-33 LOX/kerosene engine	Teaming arrangement for utilization of NK-33 family of engines in the U.S. market.	NPO Trud, Samara	Aerojet	Seeking U.S. government funding for engine validation tests.
D-57 LOX/LH engine	Teaming arrangement to improve, market, and co-produce engine for possible use in single-stage-to-orbit (SSTO) sub-scale demonstrator, or in high-performance upper stage for existing launch vehicles.	Lyulka Engine Design Bureau	Aerojet	Seeking U.S. government funding for engine validation tests.
RD-0120 LOX/LH engine	Teaming arrangement for possible modification for either bi-propellant or tri-propellant cycle, to demonstrate SSTO launch-propulsion system.	Chemiautomatics Design-Development Bureau (CADB)	Aerojet	Seeking U.S. government funding for engine validation tests.
Zenit launch services	Proposed TAA under which NPO Yuzhnoye and RSC Energia would supply the vehicle, integration would occur in the United States, and launch would take place "from international waters" from a semi-submersible platform built by Kvaerner (*Space News*, Aug. 8, 1994).	NPO Yuzhnoye (Ukraine); RSC Energia; Kvaerner Shipbuilders (Norway)	Boeing Defense and Space Group	Pending U.S. government licensing of TAA and decision to proceed.
Crystal growth experiment	Commercial contract for flight of Boeing crystal growth experiment on Mir, with return to Earth by Raduga small reentry capsule.	RSC Energia	Boeing Defense and Space Group	Experiment flew successfully in 1994.

Environmental control and life-support systems (ECLS) for spacecraft	Contract research and joint feasibility studies preparatory to International Space Station.	Scientific Research Institute of Chemical Engineering (NIICHIMASH); Institute of Biomedical Problems, Ministry of Health	Boeing Defense and Space Group	Boeing-purchased Mir flight hardware being evaluated at Boeing facility in Huntsville, AL.
Lunar-surface operations using a modified Phobos lander	"Profit-sharing" and marketing agreement.	Khrunichev Enterprise.(ZIKh); NPO Lavochkin; Zvezda	International Space Enterprises (ISE), San Diego, CA	ISE says joint development of lunar landers proceeding toward early 1998 launch.
Interactive audio-video links between Mir and U.S. classrooms	Cooperative venture.	RSC Energia	ISE	Ongoing.
Worldwide marketing of carbon- and graphite-based materials	Kaiser NIIGrafit, San Leandro, CA.	Scientific Research Institute of Graphite (NIIGrafit)	Kaiser Aerospace and Electronics Corp.	Ongoing.
Worldwide marketing of advanced composite and metallic materials	Kaiser VIAM, San Leandro, CA.	All-Russia Institute of Aviation Materials (VIAM)	Kaiser Aerospace and Electronics Corp.	Ongoing.
Space launch services utilizing the Proton from Baikonur	LKE International Inc.	Khrunichev Enterprise; RSC Energia	Lockheed Commercial Space Co., Inc.	Reported to have orders or options for 15 launches and expectations of eight firm orders by end of 1995.
Specialty metals for aerospace, including beryllium	Joint venture with exclusive rights to market all ULBA beryllium products in North America and Europe.	Ulbinskiy Metallurgical Production Combine (ULBA), Kazakhstan	Loral Corp., NY; Concord Group, CO	Ongoing.
Launch-vehicle activities	Memorandum of Agreement (MOA) on "several possible areas of cooperation in launch vehicle activities, including launch vehicle stages, ground support systems and system components."	Central Specialized Design Bureau and PROGRESS Factory, Samara, Russia	McDonnell Douglas Corp.	Active.
Unpressurized composite structures and various composite fabrication techniques and materials	MOA.	Central Research Institute for Special Machine Building, Khotkovo, Moscow Region, Russia	McDonnell Douglas Corp.	Active.

Private Sector United States-Russia Cooperative Projects (Cont'd)

Title/description	Business entity and nature of relationship	Russian entity	U.S. entity	Status
Cooperative planetary rover development	Cooperative venture	Space Research Institute (IKI), Russian Academy of Sciences; Babakin Research and Design Center (Lavochkin NPO); Scientific Research Institute of Transportation Machinery (VNIITransmash)	McDonnell Douglas Corp.; Planetary Society; Brown University	Active.
To examine and perhaps utilize Russian expertise in materials, advanced mathematics, space systems, and extended human flight	Agreement to cooperate on a series of space technology research projects	RSA/Mechanical Engineering Research Institute (IMASH), Moscow	McDonnell Douglas Government Aerospace	Active; joint centers established in Moscow and Huntington Beach, CA.
RD-170/180 engines	Marketing of RD-170 and RD-180 variant (to be developed 8/93) to U.S. government and ELV manufacturers	NPO Energomash	Pratt & Whitney Government Engines and Space Propulsion, West Palm Beach, FL	Contract signed with NASA Marshall Space Flight Center for tri-propellant engine technology development.
Docking hardware procurement	Rockwell purchasing hardware, spares and technical support for the APAS androgynous docking adapter.	RSC Energia	Rockwell Space Systems Division	Docking assembly, including APAS, delivered to NASA Kennedy Space Center in late 1994.

Stationary plasma thrusters for Western communications satellites	International Space Technology, Inc., development and marketing joint venture (later joined by SEP/France)	Fakel Experimental Design Bureau; Research Institute of Applied Mechanics and Electrodynamics, Moscow	Space Systems Loral, CA	Company funding R&D to develop and qualify power processing units utilizing Western electronics, to qualify Russian thrusters to Western spacecraft requirements, and to life test the entire system.
Resurs data	Digitization and marketing of 2m-resolution data from the Resurs series of Russian photographic satellites	Priroda Center	WorldMap International, Ltd.	Delivering digitized data to customers; no 1994 Resurs flights, but next mission expected in March 1995.

APAS = Androgynous Docking Adaptor (Russian acronym)
ELV = Expendable Launch Vehicle
LH = Liquid Hydrogen

LKE = Lockheed Khrunichev Energia International
LOX = Liquid Oxygen
NPO = Scientific Production Organization

RSC = Russian Science Corporation
SEP = Societe Europeene de Propulsion
TAA = Technical Assistance Agreement

Appendix D:
Space Cooperation with the Soviet Union (Russia): A French Point of View[1]

INTRODUCTION

Space cooperation between France and the Soviet Union (Russia after 1991) has been a very special venture for two countries belonging to opposing Cold War alliances. This relationship began during the very tense decade of the 1960s, when the space race between the United States and the Soviet Union was at its height. It was not a short-term involvement by the two nations, but an endeavor that lasted a quarter of a century (until the fall of the Soviet Union) and is still going on with Russia, though in a very different spirit. Tens of laboratories and hundreds of scientists and engineers of both countries participated, and France took part in some of the most important Soviet space missions, including interplanetary flights, space station activities, and advanced astrophysics missions (see table D-1).

This highly visible East-West technological collaboration was a unique phenomenon until the first half of the 1980s, when the Soviet space program began to open itself more broadly to Western countries. The only larger cooperative achievement has been the Apollo-Soyuz rendezvous of 1975, which was not followed by a sustained collaboration between the two superpowers.

The Cold War is over, and what has been a very special relationship is now part of an increasingly global space-cooperation environment. Russia is cooperating more and more with the multinational European Space Agency and has joined the partners of the International Space Station (ISS) project (Canada, Europe, Japan, and the United States). A very large Russian-American preparatory program to the ISS is under way and will include many rendezvous between the U.S. Space Shuttle and the Russian Mir Space Station.

In this new context, what can be learned from the long French-Soviet collaboration? Can the lessons learned during a quarter of a century of common activities be significant and useful for the future? Before these questions are addressed, it is useful to recognize that:

- the Russian political and economical system has changed, but the people in the space community and the technical culture of the Russian space industry have not really changed,

---

[1] This appendix was written for this report by Alain Dupas of the University of Paris and the French Space Agency.

TABLE D–1: Milestones of Space Cooperation Between France and the Soviet Union (Russia After 1991)[1]	
1968	Araks experiments (artificial aurora borealis created by sounding rockets)
1970	French laser reflector on the Moon rover Lunakhod–1
1971	Stereo–1 experiment on Mars–3 (solar raodioastronomy)
1971	Aureol–1 satellite with Arcad–1 experiment (gamma astronomy)
1972	Sret–1 technology satellite (piggyback launched on a Soviet rocket)
1977	French satellite Signe–3 (gamma astronomy) launched by a Soviet rocket
1982	First French human flight—Jean-Loup Chretien (PVH mission) aboard Salyut–7
1984–5	Flights of Vega–1/2 space probes (with releases of balloons in Venus atmosphere and encounters with Halley's comet)
1988	Third French human flight (Jean-Loup Chretien, Aragatz mission), including an EVA (the second French spaceflight was conducted aboard NASA's Space Shuttle)
1988	Phobos flights toward Mars
1989	Launch of Granat satellite with French gamma telescope Sigma
1990	Launch of Gamma satellite
1992	Fourth French human flight (Michel Tognini, Antares mission)
1993	Fifth French human flight (Jean-Pierre Haignere, Altair mission)

[1]This list does not include numerous experiments conducted on Soviet (Russian) scientific, meteorological, and recoverable satellites.
EVA = Extra Vehicluar Activity; NASA = National Aeronautics and Space Administration.

SOURCE: Alain Dupas, 1995.

- for a long time French-Soviet space cooperation was the main window of contact between the Russian space community and the Western world, and
- French scientists and engineers have played an important role in introducing their Russian colleagues to the international space community, particularly in the field of planetary exploration.

THE RATIONALE FOR FRENCH-SOVIET SPACE COOPERATION

■ The Political Origin of the Cooperation

There is no doubt today that Russia's participation in the International Space Station (ISS) program had a political origin. That was also the case for the beginning of the space cooperation between France and the Soviet Union in 1966. The President of France, Charles de Gaulle, was very concerned about ensuring French strategic autonomy, although the fact that France was part of the Western alliance was very clear. He had engaged France in the development of nuclear weapons and ballistic missiles (the "Force de Frappe") and had decided that his country would leave the NATO integrated military command. Space was a small, but nevertheless significant, part of this drive toward French strategic autonomy; President de Gaulle was instrumental in the creation of the French Space Agency (CNES) in 1962 and in the development of the subsequent French national space program. In 1965, France became the third country to launch an artificial satellite on its own vehicle.

It is in this context that Soviet Minister for Foreign Affairs Andreï Gromyko proposed to President de Gaulle in Moscow on April 27, 1965, that the two countries should examine the possibility of space cooperation between them. This opening was followed on July 1, 1965, by an official memorandum given to the French ambassador in Moscow. For the Soviets, this proposal was certainly a way to establish visible links with a Western power in a politically significant field and to reduce Soviet isolation in the Cold War context. For France, it was a way to demonstrate independence from the United States and to confirm its willingness to take a special position in the East-West relationship. It was by no means a disengagement from French-American space cooperation, which was doing very well at that time—the first French scientific satellite was, in fact, orbited by an American rocket in 1965.

■ The Visit of General de Gaulle to Baikonur and the Agreement of June 30, 1966

Charles de Gaulle visited the Soviet Union again in June 1966 and was invited to travel to the then-secret Baikonur Space Center, where he attended, along with General Secretary of the Soviet Communist Party Leonid Brezhnev, on June 22, the launching of a rocket. He was the first foreigner invited to Baikonur and would be the only one for nearly a decade, until the preparation of the Apollo-Soyuz flight.

This visit was closely followed by the signing of an agreement on French-Soviet space cooperation. This was done on June 30, 1966 by French Minister for Foreign Affairs Maurice Couve de Murville and his Soviet counterpart Andrei Gromyko. The agreement stressed that:

> The Governments of [France and the Soviet Union]:
>
> ▪ recognizing the importance of the study and exploration of outer space;
>
> ▪ considering that the cooperation between France and USSR in this field will enable the extension of the cooperation between the two countries and will be an expression of the traditional friendship between French and Soviet peoples [. . .];
>
> have decided to prepare and implement a program of scientific and technical cooperation between France and USSR for the peaceful study and exploration of outer space.

■ The Converging Scientific and Technical Interest of the Two Countries

The political rationale and the very high level of support it created were essential to the beginning of the cooperation. It could not, however, have enabled, by itself, the establishment of a long-term, fruitful relationship. Converging scientific and technical interests were fundamental for that.

From that point of view, the French-Soviet space cooperation:

▪ opened a lot of unique opportunities for French scientists and engineers (more experiments, large scientific spacecraft, lunar and planetary probes, recoverable payloads, manned spacecraft) complementary to national, European, and American opportunities;

▪ enabled the Soviet space community to improve the scientific value of its satellites and space probes by accommodating French experiments using advanced technologies; and

▪ enabled the Soviet scientific space community to have better contacts with the French and, through them, the Western space science community.

THE WORKING OF FRENCH-SOVIET (RUSSIAN) SPACE COOPERATION

■ Reliance on Simple Procedures

The 1966 agreement was (and still is in many ways) the basis of a very long and successful working relationship that relied on very simple procedures:

▪ Projects were approved at a yearly meeting of the French-Soviet (Russian) space cooperation committee, alternatively in France and the Soviet Union (Russia); this process is still going on.

▪ The principle was (and still is, mainly) "no exchange of funds." Each party pays for its own expenses and scientific results are shared; the only exceptions are the human spaceflights, where a participation fee is paid to the Soviet (Russian) partner.

■ The Learning Process

Some difficulties have been encountered in learning to work with the Soviets. The main issues were:

▪ meeting the right counterparts—at the beginning, contacts were organized by the Intercosmos Council (a body of the Academy of

Sciences) and did not involve the space industry, which was surrounded by secrecy,

- gaining access to industrial and launch facilities, and
- knowing the exact status of a project.

For the French specialists, working with the Soviets in the second half of the 1960s was really a "cultural shock," as can be heard in a comment by Jean-Pierre Causse, former head of the CNES Technical Center, about a common satellite project at the end of the 1960s:

> [Our experience] was based on the cooperation with NASA (FR-1 satellite and other projects), which was very open. [In the Soviet Union] everything was fuzzy. Our Soviet counterparts were extremely cautious, even if they showed a lot of good will. The academic-diplomatic procedures involved heavy, formal, and infrequent meetings, and the work progressed slowly.

With a lot of patience and good will, the situation has slowly improved. Major progress occurred during the 1980s with the involvement of a new organization representing the space industry, Glavkosmos; there was more direct access to Soviet space hardware, and much more access to Soviet space facilities. These improvements have continued with the transition from the Soviet Union to Russia and the creation of the Russian Space Agency (RSA), which works very much like NASA or CNES.

However, the collapse of the Soviet Union has created new problems that were totally unknown before: budgetary, programmatic, and procurement difficulties for the Russian partner, which no longer benefits from the high priority it had in the past and suffers from the general degradation of the economy; this is particularly true for scientific projects.

▌ The Continuity of Political Involvement

The regularly renewed support of the political actors at the highest level has been an important factor in the continuation and progress of the space cooperation between the Soviet Union (Russia)

and France. The following events have been important:

- In April 1979, Brezhnev himself proposed to French President Giscard d'Estaing the conduct of a manned spaceflight aboard Salyut-7.
- In 1985, a Mitterand-Brezhnev meeting gave the two leaders the opportunity to agree on a second joint spaceflight (Mitterand attended the launching).
- In July 1989, the principle of a long-term (10-year) agreement on manned spaceflight was approved by Minister Paul Quiles and Soviet Vice President Lev Voronin; the agreement was signed by CNES, RSA, and NPO Energia in December 1989. Four flights were planned (in 1993, 1996, 1998, and 2000); the first one has already been completed.

The French-Soviet space cooperation has survived the many political crises of the Cold War, such as the invasion of Afghanistan in 1979, which happened while preparations for the first French-Soviet human spaceflight were under way. In that case, the French government decided to go on with the project but to put the focus on the technical side of the flight and to give it a very low political profile.

▌ The Reliability of the Soviet (Russian) Partner

The Soviets (Russians) were extremely reliable partners until the end of the 1980s. No project that had been started within the cooperative framework had been canceled by the Soviet side (a few were canceled by France). The first failure was encountered in 1988 with the Phobos project (although French scientists still obtained very good results).

The situation has recently changed due to the very large difficulties encountered by the Russian Academy of Sciences in funding and supporting scientific space projects. The cancellation of the Mars '94 flight and the difficult preparation of the Mars '96 mission are consequences of this degradation.

Up to now, cooperative human spaceflights have always been conducted exactly according to the schedule fixed months before the launching. Some procurement problems have manifested themselves recently, but human missions seem to be relatively protected from the degradation of the general economical situation in Russia. The question is, however, how long can this last?

CONCLUSION

The Soviets (Russians) have been very reliable partners in space cooperation with their main past partner, France. The main ingredients responsible for the continuous success of this relationship over more than 25 years seem to be:

- strong and frequently renewed political support at the highest level,
- strong common scientific and technical interest,
- a long-term commitment able to survive political (and technical, if they arise) crises, and
- a lot of patience and good will to deal with the different social and technical cultures.

Could these recipes work for very large cooperative efforts such as the International Space Station program?

Index

☆ U.S. GOVERNMENT PRINTING OFFICE: 1995 — 387-789 / 37406

Superintendent of Documents **Publications** Order Form

P3
Telephone orders (202) 512-1800
(The best time to call is between 8-9 a.m. EST.)
To fax your orders (202) 512-2250

Order Processing Code:
*7662

Charge your order. It's Easy!

☐ **YES**, please send me the following:

_____ copies of *U.S.-Russian Cooperation in Space (144 pages)*, S/N 052-003-01410-6 at $10.00 each.

The total cost of my order is $_____. International customers please add 25%. Prices include regular domestic postage and handling and are subject to change.

_____
(Company or Personal Name)   (Please type or print)

_____
(Additional address/attention line)

_____
(Street address)

_____
(City, State, ZIP Code)

_____
(Daytime phone including area code)

_____
(Purchase Order No.)

Please Choose Method of Payment:

☐ Check Payable to the Superintendent of Documents

☐ GPO Deposit Account ☐☐☐☐☐☐☐ — ☐

☐ VISA or MasterCard Account

☐☐☐☐☐☐☐☐☐☐☐☐☐☐☐☐☐☐☐☐☐

☐☐☐☐ (Credit card expiration date)

Thank you for your order!

_____
(Authorizing Signature)   (4/95)

YES NO
May we make your name/address available to other mailers? ☐ ☐

Mail To: New Orders, Superintendent of Documents, P.O. Box 371954, Pittsburgh, PA 15250-7954

THIS FORM MAY BE PHOTOCOPIED

Superintendent of Documents **Publications** Order Form

P3
Telephone orders (202) 512-1800
(The best time to call is between 8-9 a.m. EST.)
To fax your orders (202) 512-2250

Order Processing Code:
*7662

Charge your order. It's Easy!

☐ **YES**, please send me the following:

_____ copies of *U.S.-Russian Cooperation in Space (144 pages)*, S/N 052-003-01410-6 at $10.00 each.

The total cost of my order is $_____. International customers please add 25%. Prices include regular domestic postage and handling and are subject to change.

_____
(Company or Personal Name)   (Please type or print)

_____
(Additional address/attention line)

_____
(Street address)

_____
(City, State, ZIP Code)

_____
(Daytime phone including area code)

_____
(Purchase Order No.)

Please Choose Method of Payment:

☐ Check Payable to the Superintendent of Documents

☐ GPO Deposit Account ☐☐☐☐☐☐☐ — ☐

☐ VISA or MasterCard Account

☐☐☐☐☐☐☐☐☐☐☐☐☐☐☐☐☐☐☐☐☐

☐☐☐☐ (Credit card expiration date)

Thank you for your order!

_____
(Authorizing Signature)   (4/95)

YES NO
May we make your name/address available to other mailers? ☐ ☐

Mail To: New Orders, Superintendent of Documents, P.O. Box 371954, Pittsburgh, PA 15250-7954

THIS FORM MAY BE PHOTOCOPIED